D0746008

WATER RESOURCES MONOGRAPH 1

# SYNTHETIC

# STREAMFLOWS

MYRON B FIERING

BARBARA B. JACKSON

AMERICAN GEOPHYSICAL UNION
WASHINGTON, D. C.
1971

*Water Resources Monograph Series*

S Y N T H E T I C   S T R E A M F L O W S

MYRON B FIERING AND BARBARA B. JACKSON

Copyright  © 1971 by the
American Geophysical Union
Suite 435, 2100 Pennsylvania Avenue, N.W.
Washington, D.C. 20037

Library of Congress Catalog No.  77-172418
Standard Book No.  0-87590-300-2

List Price, $2.00

Printed by
McGREGOR & WERNER INC.
Washington, D.C.

# FOREWORD

As a result of expanding population and economic pressures throughout the world, interest in water resources has increased greatly in the past decade. This interest has stimulated an expansion of investigational facilities and programs by universities and private and government organizations. There have been rapid advances in data collection, modeling of hydrologic processes, and development planning and management of water resource systems.

How to disseminate the results of this research rapidly concerns both the researcher and the user. To some extent, tutorial and 'state-of-the-art' publications may accomplish this goal by offering critical and well-articulated presentations in depth. Such publications need to emphasize the application of the research results rather than the esoteric nature of the research itself.

This year the American Geophysical Union begins publication of a new Water Resources Monograph Series with the intention of playing this publication role. *Synthetic Streamflows*, the first monograph in this new series, is also the first of three monographs designated as U.S. contributions to the International Hydrological Decade.

As far as possible, each monograph will be self-contained and will deal with a specific technique of analysis. The monograph series will offer an opportunity for critical review and implementation of recent research results by those engaged in the day-to-day problems of planning and managing water resource systems.

The preparation and publication of this series is supᵗed in part by funds provided by the U.S. Department of

iii

the Interior as authorized under the Water Resources Research Act of 1964, as amended, and in part by a grant from
Resources for the Future, Inc.

N. C. MATALAS
*Chairman, Editorial Board*
*Water Resources Monograph Series*

# PREFACE

This monograph contains virtually no mathematical de-
rivations and little statistical theory, but a large number of formulas, examples, and their phenomenological jus
tification. Moreover, many references are cited so that
interested readers can locate some of the missing details
and study them from their source documents. The purpose
of this monograph is to summarize many of the proposals
for generating synthetic streamflows, to present numerical
calculations that show the step-by-step calculations, and
to offer proposals for their application in a variety of
hydrologic engineering problems. This monograph is the
first of a series of monographs emphasizing procedural aspects of hydrologic analysis.

In this spirit we provide a few computational diffi-
culties that might typically be encountered, the tables o
random numbers actually used in the computations, and tab
ular results that do not sweep under the rug all the numerical detail. This is not a research document but
rather a manual of practice for those actively engaged in
hydrologic design. The authors were aided materially by
the editors' comments but of course bear the responsibility for any remaining errors. Mrs. Lena Young typed the
drafts and the manuscript, and Mr. William Minty prepared
the drawings; these two have worked closely with the sen-
ior author on many previous projects, and their expert
professional help is, as always, much appreciated.

MYRON B FIERING
*Harvard University*
*Cambridge, Massachusetts*

BARBARA B. JACKSON
*Harvard University*
*Cambridge, Massachusetts*

# CONTENTS

# 1  TOOL OF OPERATIONAL HYDROLOGY

## Introduction and Background

Synthesized streamflows have become an important tool to the
water resource planner because, used in conjunction with computer
simulation, they allow him to evaluate proposed system designs more
thoroughly and in a more statistically sophisticated manner than was
possible with previously available methods. Simulation is certainly
not a new idea, and the desire to evaluate proposed designs is even
older; what operational hydrology and synthetic streamflows add to
the planning process is the capacity for increased scope and sensi-
tivity of evaluations.

Early civil engineers realized of course that the flow patterns
of different streams vary considerably and that the history of flows
in a particular stream provides a very valuable clue to the future
behavior of that stream. Hence Rippl (1883) devised the familiar
mass curve analysis to investigate the storage capacity required to
provide a desired pattern of drafts despite inflow fluctuations.
Mass curve analysis examines the historical or any other specific
record to determine the minimal amount of storage required to smooth
out fluctuations by meeting given target releases, all subject to
the given flow pattern.

Many design studies use mass curve analysis to study storage requirements, both for within-year and for over-year storage. Similarly, the historical record for a stream can be (and has been) used
in examining other aspects of a water system; e.g., the frequency
and the persistence of periods of high flows in a historical record
provide an indication of flooding patterns. An engineer might
design a system to provide storage and flood control, for example,
and then perform routing studies by subjecting the proposed design

1

to the historical flow sequence and thereby testing the design's appropriateness.  Such use of the historical record alone, however, involves several potentially serious problems.

1.  The historical record of flows for any given stream is apt to be quite short, frequently less than 25 years.  The exact pattern of flows during that historical period is extremely unlikely to recur during that period in which the proposed system would be operative (i.e., the economic life of the system).  Furthermore the recorded values of high flow, low flow, and other characteristics of the record are not at all likely to be maintained during the economic life of the system.  In all cases, planners would certainly agree that the worst flood (or drought) on record is not the worst possible flood (or drought).  Also the historical record may not be long enough to provide the planner with as much data as he needs.  For example, the record may not be as long as the proposed economic life of a dam or of a system of structures for flood control.

2.  An important fact is that in evaluating proposed designs this use of the historical record alone gives no idea of the risks involved.  If a storage is designed to meet target drafts despite droughts, the engineer wants some idea of the likelihood that the dam will run dry during its economic life.  Asking this question is not equivalent to asking if the dam would have run dry during the period of record.  The planner might like to know, for example, that there is a 95% chance that the proposed dam will suffice for its intended use.  In other cases, it might be that a proposed water resource system would frequently fail to meet discharge or quality standards but that such failures would be minor.  An example might be that the dissolved oxygen level in a stream would usually fall very slightly below some desired level, but would always be near that level.  Operational hydrology can help to provide the engineer with estimates of the expected frequency and severity of failures.

### Operational Hydrology

To determine statistics such as the expected frequency of droughts (of a given severity) in a particular stream, the planner may use either a very large sample of flows from the stream in question or the exact probability distribution of flows in the stream. Given the distribution of flows, he could determine analytically the expected frequency through manipulation of the flow distribution function or, if that approach proved intractable, he could use Monte Carlo techniques to estimate this (or any other relevant) statistic. (Monte Carlo techniques employ random sampling to estimate quantities. For example, suppose we wish to estimate the integral of a function over the interval $[0, 1]$ and we know that the function's value is between 0 and 2 over that interval. We consider the rectangle of area 2 with vertices $(0, 0)$, $(1, 0)$, $(0, 2)$, $(1, 2)$; this rectangle encloses the function. Next we take a (large) random sample of points in the rectangle and we determine for each point whether it is above or below the function. Assume that a fraction $p$ of the points lie on or below the curve. Then $p$ is used as an estimate of the part of the area of the rectangle (2 units) that lies below the curve; $2p$ is an estimate of the desired integral.)

With a sufficiently large sample of streamflows, the planner would calculate the sample value of the statistic, for example, the proportion of years with droughts of the given severity. Several assumptions are implicit in such procedures. One is that there is some probabilistic mechanism underlying the physical generation of the streamflows and that this mechanism is sufficiently stable over the period studied that the process can be considered stationary. (Intuitively, a stationary process is one whose main characteristics (or parameters) remain constant over time. A process is called weakly stationary if its average value and average spread remain constant over time; strong stationarity means that all statistical moments are constant over time (see the two sections in chapter 2 on the calculation of the statistics of the flow distributions).) The planner also assumes that his sample is representative. Therefore sample statistics are valid approximations of the corresponding characteristics of the underlying process.

Hydrology is unable to provide either the exact probability distributions of flows or long records of past flows, unfortunately.

Hydrologists have devised some causal models of flow generation (e.g., Crawford and Linsley, 1962), but there is not yet a conclusive model of streamflows, let alone one that is capable of predicting future streamflows.  As noted in the previous section, existing streamflow records are not sufficiently extensive to provide estimates of many important statistics.  One answer to this dilemma is to generate records which, while neither actual historical records nor predictions of future flows, are close enough to possible (but not observed) historical records that they may be used to determine, within defined statistical errors, the several quantities of interest.  Operational hydrology serves this function.  The phrase close enough is purposely vague; its meaning will vary with the planner's needs as reflected by system response, and much of this monograph concerns what kinds of closeness are possible, desirable, or important in different contexts.

Such generated or synthetic streamflows are extremely useful planning tools.  It is important to realize that they are generated by statistical methods that do not pretend to provide causal models for actual flows.  Instead, justification for the use of such tools is entirely pragmatic.  Given a synthetic trace of 100 different flow sequences, each of which is equally likely to be the actual sequence observed during the economic life of the structure, the planner can perform routing or other design studies on all these sequences and hence obtain quantitative statistical measures both of the anticipated performance and of risks.  It would be intellectually and aesthetically satisfying if a complete, theoretically tidy model of river flows were available.  At present, however, only approximate models are available.

Nevertheless, such approximate models are sufficiently realistic that their use improves the planning process significantly.  Because operational hydrology provides large numbers of flows for use in evaluating storages, much calculation is involved (and hence electronic computers are needed).  The relation here is actually even

more basic: modern electronic computers make the large-scale use of
operational hydrology possible.  Early attempts were made to use
manually generated streamflows in hydrological studies (Sudler,
1927), but  the tremendous amount of calculation required in a thorough study is feasible only on a computer.

Hydrologists and meteorologists have provided new insights into
the characteristics and mechanisms of flows.  They have, for example, recognized that there is an important tendency toward persistence in successive streamflows; low flows tend to be followed by
other low flows, high flows tend to be followed by high flows.  Such
information is valuable for operational hydrology because users can
generate flows that retain the statistical characteristics of observed flows and they can judge alternate generation schemes by the
extent to which the generated flows display the expected statistical
characteristics.  Hydrology and meteorology also indicate which phenomena and processes in a water basin tend to be strongly correlated
with streamflows.  New statistical estimating techniques enable the
engineer to use regional data on these phenomena to refine estimates
of moments and thereby to improve flow generation for a given stream
or even to generate flows for streams for which there is no historical flow record and, hence, for which flow generation would otherwise be impossible.  Such regional techniques are discussed later.

The general area of operations research or systems analysis has
provided operational hydrology and simulation with the concept of a
system objective function, a function for assigning a numerical index of value to a design so that alternate designs can be evaluated,
compared, and ranked in a systematic manner.  (Recent theory has
dealt with problems of multiattributed objective functions, for
which a vector or array of index values is required, and with the
relationships among the system constraints and objectives (Raiffa,
1968).)  The use of economics and decision theory has provided fun-
damental insights into the basis for such evaluation of designs;
their contribution is discussed below.

### Role of Decision Theory

Synthetic streamflows represent a prescriptive, not a descriptive, modeling process. Because their justification is not causal but is a tool in planning and evaluation, it is appropriate to consider briefly the general process of design evaluation through simulation. Thus we assume for now that appropriate synthetic flow traces are available; the specific schemes for generating such traces are discussed later. The problem of selecting a reservoir size for a storage system intended to withstand droughts is used as a prototype. However, the concepts of objective function, costs, benefits, and risks that are used here are certainly also pertinent in other design situations. Assume, then, that we are testing the performance of a reservoir of size 10 units on a stream with a mean (average) annual flow of 6 units and with a target draft of 6 units. The reservoir is not being designed in this analysis. Either the reservoir already exists or this analysis is part of a larger design study for which many storage sizes and operating policies are tested. The reservoir management scheme is the normal operating policy shown in Figure 1. It is further assumed that the total available supply, the abscissa in Figure 1, is known at the start of the time period (e.g., year). Thus if fewer than 6 units are available, all the water is released; if total available water (current storage plus current inflow) is more than 16 units (reservoir capacity plus target draft), the reservoir is filled and all additional water is released; in all intermediate states exactly 6 units are released and the reservoir is left at some intermediate level. The simulation program uses synthesized streamflows for inputs to the reservoir; it uses the normal operating policy as a rule in determining the outflows (and hence the storage) in successive periods. Hypothetical sets of inflows, outflows, and storage levels (assumed to be abstracted from the middle of a simulation run) are given in Table 1. The flows are simply subject to a traditional routing procedure.

The next question is how to interpret the results in Table 1. The analyst can run many sets of streamflows through such a routing

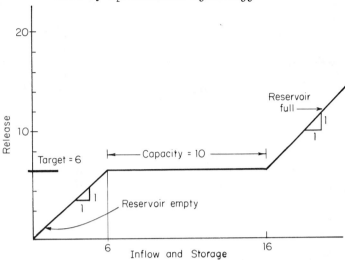

Fig. 1. Reservoir operating rule.

procedure. He might, for example, have the results for 100 sets of data, each as long as the proposed economic life of the structure. Operations research and decision theory now help answer the question: What next? The analyst must first have some objective or standard for judging various designs. An objective function considers both the benefits provided by a structure and the costs related to its construction; the benefits and costs must be expressed in comparable quantitative terms. One objective is to maximize expected net benefits, the difference between expected benefits and costs. A second objective is to maximize the ratio of expected benefits to costs; this objective function measures the return in benefits per dollar invested. A third objective is to minimize costs subject to the constraint that benefits reach a prescribed level; a fourth objective is to maximize benefits subject to certain cost constraints. These different objectives do not generally suggest the same optimal design, and only one objective can be used. The selection is frequently influenced by political and social processes.

Determination of the cost portion of an objective function is generally the easier part of the analysis because it is natural to consider most costs in quantitative (probably dollar) terms. It is usually harder to select an appropriate measure for benefits. The

TABLE 1.   Routing Study, 20-Year Record

| Period | Initial Storage | Inflow | Release | Final Storage | Deficit | Surplus Release |
|--------|-----------------|--------|---------|---------------|---------|-----------------|
| ... | ... | ... | ... | ... | ... | ... |
| ... | ... | ... | ... | ... | ... | ... |
| ... | ... | ... | ... | 9 | ... | ... |
| 1 | 9 | 12 | 11 | 10 | 0 | 5 |
| 2 | 10 | 11 | 11 | 10 | 0 | 5 |
| 3 | 10 | 12 | 12 | 10 | 0 | 6 |
| 4 | 10 | 5 | 6 | 9 | 0 | 0 |
| 5 | 9 | 4 | 6 | 7 | 0 | 0 |
| 6 | 7 | 2 | 6 | 3 | 0 | 0 |
| 7 | 3 | 1 | 4 | 0 | 2 | 0 |
| 8 | 0 | 3 | 3 | 0 | 3 | 0 |
| 9 | 0 | 5 | 5 | 0 | 1 | 0 |
| 10 | 0 | 8 | 6 | 2 | 0 | 0 |
| 11 | 2 | 5 | 6 | 1 | 0 | 0 |
| 12 | 1 | 4 | 5 | 0 | 1 | 0 |
| 13 | 0 | 5 | 5 | 0 | 1 | 0 |
| 14 | 0 | 7 | 6 | 1 | 0 | 0 |
| 15 | 1 | 8 | 6 | 3 | 0 | 0 |
| 16 | 3 | 2 | 5 | 0 | 1 | 0 |
| 17 | 0 | 2 | 2 | 0 | 4 | 0 |
| 18 | 0 | 1 | 1 | 0 | 5 | 0 |
| 19 | 0 | 7 | 6 | 1 | 0 | 0 |
| 20 | 1 | 11 | 6 | 6 | 0 | 0 |
| ... | | | | | 18 | |

Capacity = 10; target release = 6.

planner can consider such diverse benefits as the output or physical
performance of the structure as opposed to the extent to which the
structure (and its construction) encourage income redistribution in
the population.   To simplify this discussion of evaluating a reservoir design, we consider performance indices only.

As a first approximation, the analyst might use as his measure of
performance for the problem at hand the proportion of years in
which the target draft is successfully met.   In the 20-year sequence
shown in Table 1, the draft is 6 units (or more) in 12 years and so
the proportion of adequate drafts is 12/20 = 0.60.   This proposed
index of reliability is not entirely inappropriate, but it is not
altogether satisfactory.   The planner would want to know how bad the

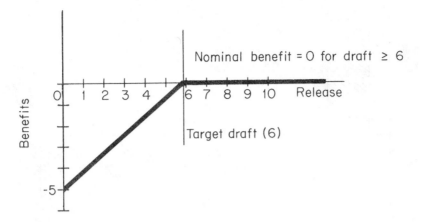

Fig. 2.   Linear penalty function.

deficits are, for it might be that a release of 5 units is not nearly as bad as a release of 1 unit.  As a second approximation to a
performance index the analyst might use the sum of deficits in various years.  This parameter is actually a measure of badness, so the
planner would use its negative as an index.  In the 20-year sequence
shown, the deficits are 6 - 4 = 2 in the seventh year, 3 in the
eighth, 1 in the ninth, 1 in the twelfth, 1 in the thirteenth, 1 in
the sixteenth, 4 in the seventeenth, and 5 in the eighteenth or a
total of 18 units of deficit; hence the proposed benefits measure
is -18.  This performance function is really a piecewise linear one
that penalizes deficits linearly whereas it ignores surpluses; it
measures the area below the time axis in a plot of surplus releases
versus time.  The function is shown in Figure 2.  Figure 3 shows
surplus release versus time for the data in Table 1.

    For some planning purposes the second performance function above
is a definite improvement over the first but is not ideal.  If it
happens that a deficit of 2 units is considerably more than twice as
damaging as a deficit of 1 unit, the analyst might next consider a
nonlinear function for penalizing deficits, such as the quadratic
penalty function in Figure 4.  (As suggested by Leo R. Beard, in a
variety of work, the U.S. Army Corps of Engineers uses a quadratic
loss or penalty function (Close et al., 1970).)  Next, he might decide that water above the target level of 6 units per year is

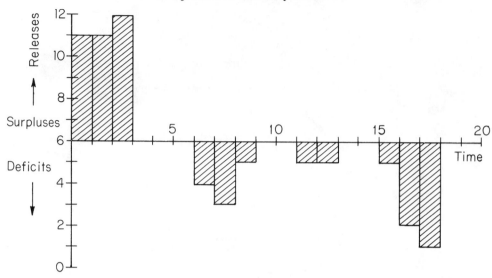

Fig. 3.   Surplus and deficit releases for the data of Table 1.

valuable, although not as valuable as the first 6 units each year.
Consequently, he might adopt a two-part linear performance function
(Thomas, 1959; Bryant, 1961) as shown in Figure 5.  Finally, the
analyst might realize that the actual harm caused by a draft deficit, whatever its size, in a given year is partly determined by the
draft patterns in previous and following years.  An isolated deficit
year may be bad, but a series of such deficit years can be devastat-
ing.  Therefore a good performance index might include measures of
persistence of losses.

Many other refinements of performance indices are available to a
planner trying to determine the benefits of alternative designs.  In
an actual planning situation, an analyst tries to find a particular
evaluator that is sophisticated enough to measure what he considers
to be the important aspects of performance but yet simple enough to
be useful computationally.  Usually the analyst is forced to accept
a compromise measure.

The general objectives of simulation studies largely determine
the appropriate way to conduct these studies.  In the storage example above, if the objective is to determine the smallest dam that
will provide a given performance level (as measured by a carefully
chosen performance index), the analyst would run simulations for

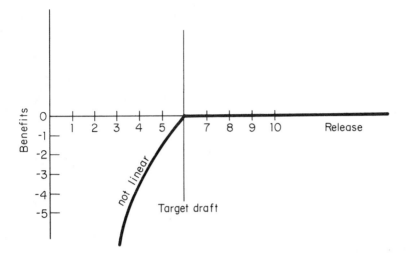

Fig. 4.   Quadratic penalty function.

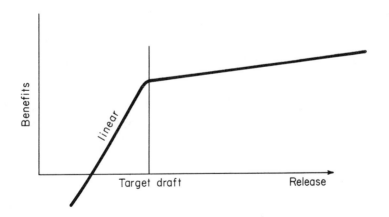

Fig. 5.   Two-part linear benefit function.

various dam sizes and determine the resultant performance indices
and would then select the smallest size that gave a performance index at least as good as the specified value.  On the other hand, if
the analyst is constrained to use a given dam size (because the dam
already exists), he might investigate the performance of the dam for
different target drafts or even for operating policies other than
the normal one described here.

   This brief discussion suggests the general environment for the
use of synthetic streamflows in simulation; it is certainly not complete.  For further information on simulation, decision theory, and
cost-benefit analysis, the reader is referred to Tocher (1963),
Maass et al. (1962), Hufschmidt and Fiering (1966), and Benjamin and
Cornell (1970).

# 2 STATISTICS OF OPERATIONAL HYDROLOGY

## Time Series and Random Numbers

To generate a sequence of values of synthetic flows for a given
stream, we consider the flows to be the results of a random process,
a process whose results change with time in a way that involves
probability (Moran, 1959). Thus we assume that we can make at least
approximate statements of the form 'the probability or chance that
the flow in a given stream next year will be less than $x$ units is
$p_1$.' We do not assume that the exact flow can be predicted; in fact,
we do not even attempt to evaluate the extent to which the actual
generating process for real flows involves deterministic laws and
the extent to which it involves chance. Rather, perhaps simply because of our ignorance, we assert that we see the generation process
as a random one, and then we attempt to model it. Of course the
historical flow sequences do not seem entirely whimsical or haphaz-
ard. At the very least, a stream that has shown a mean flow of 10
units per year over the years of record and has not undergone any
catastrophic natural or man-made changes is very unlikely to start
flowing at a sustained average rate of 20 units per year. It is
much more likely that the mean flow will remain near 10 units per
year. Moreover, if most of the recorded flows lie close to 10
units, with only rare extreme flows, we expect that with high probability the next flows will be near 10. Thus we expect the general
level of variability of the flows to be maintained. Conversely, if
the past flow pattern is very erratic, with many high flows and many
low flows, we expect a similar spread in the future. Other characteristics of past flow sequences provide clues to future flows. If
the flow this year is low, it is likely, although not certain, that
the flow next year will also be lower than average. Similarly, high

flows tend to follow high flows.  Thus the history of a stream provides valuable information about probable future flows.  Models for
generation should certainly use such information although, at the
same time, they must include a random component to reflect our in-
ability to predict future flow sequences exactly.

A set of historical or synthetic flows for a stream is a sequence
of numbers or values produced by a random process in a succession of
time intervals; such a sequence is called a time series.  In general, the $i$th member of a time series, which we write as $x_i$, is the
sum of two parts:

$$x_i = d_i + e_i \qquad\qquad (1)$$

Here $d_i$ is a deterministic part, a number determined by some exact
functional rule from the parameters and previous values of the process.  Typically $d_i$ might be a function of the mean flow, of the variability of flows (as measured by their standard deviation), and of
previous flows such as $x_{i-1}$, $x_{i-2}$.  The random component of the generating scheme is $e_i$.  It is a random number drawn or sampled from
the set of random numbers with a certain probability distribution or
pattern.  In one case $e_i$ might be drawn from what is called the uniform distribution on $[0, 1]$; in this distribution each number between 0 and 1 is just as likely to occur as each other number (Figure 6a).  In other cases, $e_i$ might be drawn from the familiar bell-
shaped normal distribution shown in Figure 6b.  The graphs in Figure
6 show variables along the horizontal axis and the frequency values
in the vertical direction.  The integral of the curve between two
numbers, such as $e_1$ and $e_2$ in Figure 6b, is the probability that a
number drawn from a population with this distribution will lie between the chosen numbers.

In setting up the deterministic part of a generation scheme for
streamflows, the analyst considers such questions as: Does the previous flow influence the current flow?  If so, how?  What about the
flow before the previous one and the one before that?  Should other
hydrologic data such as precipitation be included?  Several possible
forms for the deterministic portion are discussed below, but first

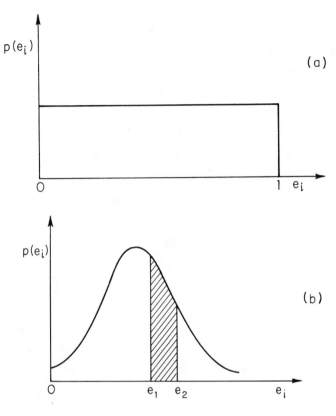

Fig. 6.   (a) Uniform distribution on
[0, 1].   (b) Normal distribution.

we consider the random component and how random numbers can be generated by computer.   Parts of a useful discussion of random number
generation (Knuth, 1969) are surveyed here briefly.

The programmer who is coding a simulation problem must have a
source of random numbers.   One obvious source is any random process
in nature; the emission of particles from radioactive material is
one example.   Such sources are not really satisfactory since the
numbers generated would have to be recorded, then read into the computer and stored in valuable memory locations.   The entire procedure
would be expensive and awkward.   Instead, computers use their own
arithmetic capabilities to generate random numbers.   Computers are
deterministic machines (barring such undesirable occurrences as power failures) in that, given identical inputs, they produce identical
results.   It seems impossible for a deterministic machine to use a

deterministic algorithm or computational scheme to generate random
numbers, and it is in fact impossible. Computers can, however, generate pseudorandom numbers, which are sequences of numbers carefully
(deterministically) constructed to maintain the important properties
of truly random sequences. The basic pseudorandom number generators
produce uniformly distributed numbers, from which numbers following
other distributions can be generated by manipulating the sequences
of uniformly distributed random numbers.

Intuitively, a sequence of numbers with a uniform distribution is
a sequence in which each element is generated randomly, without ref-
erence to other elements in the sequence, and in which each number
in the allowable range (in many cases 0 to 1) is as likely to occur
as is any other number in that range. (One procedure for generating
pseudorandom uniformly distributed numbers is the linear congruen-
tial method that operates as follows. Select some starting value
$x_0 \geq 0$, a multiplier $a \geq 0$, an increment $c \geq 0$, and a modulus (or
divisor) $m$ which is larger than $x_0$, $a$, and $c$. Use $x_0$ as the first
element in the sequence. Let $x_1 = (a\ x_0 + c)$ mod $m$ where this mod
notation means to take the quantity $a\ x_0 + c$, subtract $m$ as many
times as possible without driving the result negative, and then set
$x_1$ equal to the result. In an equivalent form $x_1$ is the remainder
when $a\ x_0 + c$ is divided by $m$. (For example, to form 15 mod 7,
start with 15 and subtract 7 to obtain 8. Since $8 \geq 7$ we can subtract 7 again to obtain 1. Since $1 \leq 7$ no more subtractions are
necessary and the result is 1. Or, since 15/7 = 2 with a remainder
of 1, the answer is 1.) At each subsequent step $x_n$ is generated
from the previous value $x_{n-1}$ by the formula $x_n = (a\ x_{n-1} + c)$ mod $m$.
Knuth discusses rules that should be applied in selecting the $a$, $m$,
and $c$ values so as to make the sequence as random as possible.)

All algorithms for generating random numbers are based on earlier
values of the generated sequence; hence they are called pseudorandom
number generators. Moreover, a sequence of numbers generated within
a computer by a deterministic algorithm must repeat. Computers use
only numbers that contain less than some set number of digits, since

within the computer there is only a fixed amount of storage space
reserved for any number.  Therefore if there are $p$ possible numbers
with that set number or fewer digits, the algorithm must repeat
after $p$ (or fewer) numbers are generated.  Knuth's algorithm and associated rules are intended as aids in making the repetition period
or cycle as long as possible.  In practice the user will not be both-
ered by repetitions because $p$ is usually an extraordinarily large
number (typically $p$ might be $2^{32}$ = 4,294,967,296).  It is this fact
that makes pseudorandom number generators usable.

Two points about uniformly distributed random numbers should be
mentioned.  First, infinite sequences from a generator are never
available; on finite computers all sequences eventually cycle.  Real-
world tests of randomness can only require that individual digits or
sets of digits (pairs, triples, and so on) be almost uniformly distributed.  Second, humans are really not very good at generating or
at detecting randomness without using mathematical tools.  People
tend not to realize how frequent strings of repetitions of the same
digit in truly random sequences are.  The set of digits 6666 should
appear just as frequently as any other quadruple of digits but most
tests of 'random' digits generated unsystematically by humans do not
contain enough such runs.  In other words, random runs do not seem
very random to humans.

Finally, we must consider the problem of generating random numbers with other distributions, particularly normally distributed
numbers.  Fortunately, it is quite easy to generate normal random
numbers from a sequence of uniformly distributed random numbers in
the range [0, 1].  We want to generate normal random numbers with
zero mean and unit standard deviation (which in this case equals the
average square of the numbers).  Such numbers are called standard
normal deviates.  They have the frequency function

$$f(t) = (2\pi)^{-1/2} \quad \exp (-1/2 \ t^2) \tag{2}$$

The central limit theorem of probability theory says that numbers
formed by taking the sums of random numbers from the uniform

distribution (or from most other distributions) are approximately
normally distributed if enough different numbers are included in each
sum; 12 will be enough numbers here.  Thus if $u_1$, $u_2$, $u_3$, ... is a
sequence of uniformly distributed random numbers,

$$t_1 = u_1 + u_2 + u_3 + \ldots + u_{12} - 6 \tag{3}$$

$$t_2 = u_{13} + u_{14} + \ldots + u_{24} - 6 \tag{4}$$
$$\vdots$$

will be approximately normally distributed with mean 0 and standard
deviation 1.  The -6 is required here to give zero mean.

   Another method for transforming rectangular $[0, 1]$ deviates into
normal $[0, 1^2]$ deviates is based on the use of two rectangularly distributed values $(u_1, u_2)$:

$$t_1 = \sqrt{\ln \frac{1}{u_1}} \, \cos \, 2\pi u_2 \tag{5a}$$

$$t_2 = \sqrt{\ln \frac{1}{u_1}} \, \sin \, 2\pi u_2 \tag{5b}$$

where $(t_1, t_2)$ are normally distributed with zero mean and unit variance.  This transform is due to Box and Muller (1958).

   Table 2 gives a set of random numbers drawn from the uniform distribution.  Table 3 gives random numbers drawn from a normal distribution with mean zero and unit standard deviation.  Normal random
deviates with this mean and variance are called standard normal deviates.  In this monograph the random components in generating equations have zero mean and unit variance.  It is simple, however, to
generate random numbers with different means and variances, and for
completeness we include the calculation.  Assume that $y_1$, $y_2$, ... is
a sequence of random numbers drawn from some distribution with mean
(average value) 0 and variance (average squared distance to the
mean) 1.  Then the sequence $z_1$, $z_2$, ... can be considered to come
from the same family of distributions (normal, for example) but with
mean $\mu$ and variance $\sigma^2$ if the sequence is formed by

$$z_i = \mu + \sigma y_i \qquad (6)$$

## Statistics of the Flow Distribution:
## Mean and Standard Deviation

We now consider the second part of specifying the time series for generating synthetic flows, that of specifying the deterministic element of the flows. To do this, we consider the important characteristics of the historical flow record and alternate ways of making the generated flows show the same characteristics.

First, we insist that the generated flows have the same average value or mean as the observed flows. If the historical record contains $n$ yearly flows, for example, the sample mean flow is

$$\bar{x} = \frac{1}{n} \sum_{i=1}^{n} x_i \qquad (7)$$

which is an estimate of the true or population mean $\mu$. Mathematically we have

$$\mu = E[\bar{x}] \qquad (8)$$

where $E[\bar{x}]$ denotes the expectation of $\bar{x}$ as $n$ tends to infinity; that is, $E[\bar{x}]$ is the limiting value (in a probability sense) of $\bar{x}$ as $n$ tends to infinity and where the $x_i$ are annual flows. There are several qualifications concerning even this very simple statistic. For one thing, the most that any generating scheme can promise about this or any other statistic is that the expected value of the same statistic of the synthetic flows is the desired value. The expected value is the average value of the statistic in an infinitely long sequence of generated flows. Since the generation algorithms are only used to form finite sequences of flows, the sample means obtained cannot be expected to equal the historical mean exactly. However, they tend to be near the historical mean, and the closeness is expected to increase (i.e., improve) with the length of the generated sequences. Abundant statistical theory considers questions of how close to their theoretical values various sample statistics

TABLE 2.  Uniformly Distributed Random Numbers

|    | 0 | 1 | 2 | 3 | 4 | 5 | 6 | 7 | 8 | 9 |
|----|------|------|------|------|------|------|------|------|------|------|
| 0  | 0.132 | 0.736 | 0.795 | 0.799 | 0.165 | 0.640 | 0.314 | 0.420 | 0.915 | 0.867 |
| 1  | 0.104 | 0.319 | 0.930 | 0.100 | 0.731 | 0.393 | 0.855 | 0.720 | 0.581 | 0.399 |
| 2  | 0.491 | 0.434 | 0.390 | 0.649 | 0.638 | 0.089 | 0.978 | 0.228 | 0.075 | 0.810 |
| 3  | 0.582 | 0.834 | 0.074 | 0.581 | 0.137 | 0.385 | 0.807 | 0.400 | 0.901 | 0.497 |
| 4  | 0.024 | 0.125 | 0.788 | 0.982 | 0.386 | 0.145 | 0.788 | 0.472 | 0.817 | 0.992 |
| 5  | 0.847 | 0.302 | 0.254 | 0.024 | 0.583 | 0.776 | 0.175 | 0.307 | 0.212 | 0.942 |
| 6  | 0.268 | 0.814 | 0.065 | 0.996 | 0.633 | 0.069 | 0.058 | 0.451 | 0.873 | 0.362 |
| 7  | 0.506 | 0.181 | 0.464 | 0.662 | 0.638 | 0.475 | 0.772 | 0.346 | 0.213 | 0.230 |
| 8  | 0.096 | 0.875 | 0.577 | 0.393 | 0.102 | 0.542 | 0.577 | 0.788 | 0.039 | 0.414 |
| 9  | 0.668 | 0.459 | 0.100 | 0.591 | 0.619 | 0.631 | 0.303 | 0.918 | 0.171 | 0.444 |
| 10 | 0.279 | 0.261 | 0.038 | 0.559 | 0.322 | 0.479 | 0.611 | 0.776 | 0.099 | 0.219 |
| 11 | 0.757 | 0.787 | 0.375 | 0.652 | 0.070 | 0.180 | 0.139 | 0.288 | 0.968 | 0.835 |
| 12 | 0.507 | 0.973 | 0.990 | 0.819 | 0.768 | 0.220 | 0.685 | 0.920 | 0.954 | 0.975 |
| 13 | 0.463 | 0.935 | 0.317 | 0.958 | 0.782 | 0.843 | 0.063 | 0.703 | 0.037 | 0.988 |
| 14 | 0.752 | 0.405 | 0.548 | 0.385 | 0.710 | 0.730 | 0.002 | 0.299 | 0.844 | 0.682 |
| 15 | 0.729 | 0.192 | 0.230 | 0.673 | 0.196 | 0.189 | 0.845 | 0.905 | 0.077 | 0.169 |
| 16 | 0.832 | 0.207 | 0.297 | 0.620 | 0.654 | 0.438 | 0.397 | 0.371 | 0.524 | 0.583 |
| 17 | 0.622 | 0.337 | 0.491 | 0.671 | 0.090 | 0.392 | 0.535 | 0.641 | 0.421 | 0.506 |
| 18 | 0.073 | 0.275 | 0.206 | 0.256 | 0.234 | 0.327 | 0.680 | 0.856 | 0.281 | 0.975 |
| 19 | 0.988 | 0.808 | 0.556 | 0.711 | 0.409 | 0.321 | 0.332 | 0.811 | 0.916 | 0.604 |

TABLE 3.  Standard Normal Random Sampling Deviates

|    | 0 | 1 | 2 | 3 | 4 | 5 | 6 | 7 | 8 | 9 |
|----|------|------|------|------|------|------|------|------|------|------|
| 0  | -0.523 | 0.611 | -0.359 | -0.393 | 0.084 | -0.931 | -0.027 | 0.798 | 1.672 | -1.077 |
| 1  | -1.536 | -0.454 | 0.071 | -2.129 | 1.525 | 0.261 | 2.319 | 0.972 | 0.767 | -2.849 |
| 2  | -0.121 | 0.968 | -1.943 | 0.581 | -0.711 | -0.060 | -0.482 | -0.746 | -0.747 | 1.254 |
| 3  | -0.542 | -0.807 | 0.168 | 0.839 | -0.756 | -0.453 | -1.912 | 0.766 | -0.890 | 0.205 |
| 4  | 0.131 | -0.859 | -1.096 | -0.785 | 0.310 | 1.314 | -0.231 | 0.029 | 1.819 | -1.602 |
| 5  | -0.234 | 0.551 | 0.743 | -0.900 | 0.435 | -2.999 | 0.212 | 0.869 | -0.716 | -0.410 |
| 6  | -1.010 | 1.347 | 0.230 | 0.009 | -1.495 | 2.145 | -1.033 | 0.729 | 0.309 | 0.920 |
| 7  | 0.273 | -0.885 | -0.016 | 0.775 | -1.740 | 0.353 | -1.519 | 0.958 | -0.448 | 2.185 |
| 8  | -0.102 | -1.111 | -0.585 | 1.461 | -0.307 | 1.489 | -0.196 | 0.506 | -0.662 | -1.175 |
| 9  | 0.368 | -0.710 | 0.407 | 0.066 | -0.617 | -0.580 | 0.107 | -2.247 | 1.616 | -1.060 |
| 10 | -1.762 | 1.382 | 1.142 | -2.056 | -0.400 | -1.701 | -0.914 | -1.000 | -0.172 | 0.903 |
| 11 | 0.306 | -0.607 | -0.324 | 1.171 | 1.016 | -1.829 | 1.723 | -0.513 | -0.657 | 2.011 |
| 12 | -0.465 | -1.214 | -0.174 | 0.894 | 0.245 | -0.987 | -1.155 | 0.592 | -0.411 | -0.109 |
| 13 | -0.004 | -0.029 | -0.633 | 0.004 | -0.603 | 1.104 | -0.655 | 1.191 | 0.938 | -0.805 |
| 14 | 0.593 | 0.252 | -0.541 | 0.318 | 1.268 | 1.972 | 0.875 | -1.030 | -1.175 | 0.445 |
| 15 | 0.233 | 0.430 | -0.331 | -1.272 | -0.289 | -0.060 | -0.754 | 0.789 | 0.546 | 0.687 |
| 16 | 0.571 | -0.215 | -1.090 | 0.610 | -0.810 | -0.364 | -1.282 | 0.010 | 0.586 | 0.926 |
| 17 | 0.370 | 0.976 | 1.017 | 1.106 | 0.441 | -2.376 | 0.793 | 0.016 | -0.704 | 0.146 |
| 18 | -0.009 | -1.285 | -0.346 | -0.323 | 0.609 | -0.373 | 0.078 | -1.034 | 0.153 | 0.997 |
| 19 | 0.416 | -0.131 | 0.668 | 0.662 | -1.835 | 1.646 | 0.197 | 0.131 | 0.783 | 0.076 |

can reasonably be expected to lie.  In the rest of this discussion,
when we say that a given algorithm generates flows with some mean,
standard deviation, or other parameter, we imply that the expected
value of that parameter is the specified value.  The actual value of
that parameter for the particular sample or synthetic trace will be
within theoretically anticipated sampling errors of the expected or
population value.  This procedure is equivalent to citing a population mean (say μ) and constructing a set of samples whose means are
distributed around μ but for which the mean of all means tends
closer to μ as the sample size increases.

The second qualification is a particularly important one for hydrologic modeling.  Because we use the historical or sample values
of such quantities as mean flow to estimate the true or population
values of these quantities for the actual process, we must have some
idea of how good the sample estimates will be.  The answer to the
problem, unfortunately, is ambiguous and clouded by statistical un-
certainty.  One obvious rule is that small samples are not very reliable.  However, the chance of obtaining an unrepresentative sequence in a large sample is much smaller than it is in a small sample.

How large a sample is large enough for streamflow data?  Again,
the answer is uncertain.  Historical flow sequences are characterized by persistence.  Typically a low flow is more likely to be
followed by another low flow than to be followed by a high flow, and
similarly a high flow is more likely to be followed by another high
flow.  The statistical description of the phenomenon is that successive flows are positively correlated.  Such persistence reduces
the true amount of information about the population mean that is
contained in a sample of given size.  For example, consider the two
samples shown in Figures 7a and 7b.  The successive elements in the
first series do not display noticeable persistence.  The entire
sample provides a good estimate of the population mean, but so do
the elements of only the first half of the sample or those of only
the second half of the sample.  In the second series (Figure 7b)

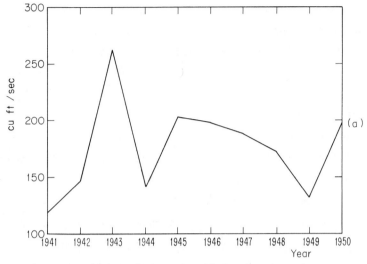

Fig. 7    Mean annual discharge in cubic feet/sec Susquehanna
          Basin : Oaks Creek at Index, N.Y.

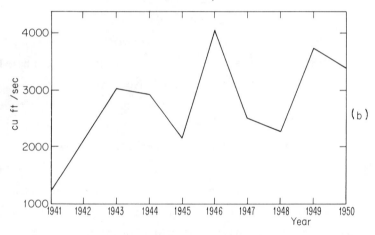

Fig. 7.  (a) Mean annual discharge in cubic feet per
  second, Susquehanna basin: Oaks Creek, Index, New York.
  (b) Mean annual discharge in cubic feet per second,
  Oostanaula River, Resaca, Georgia.

there is a high level of persistence.  The high flows are generally
bunched together as are the low flows.  The entire sample provides
a good estimate of the population mean, but the estimate obtained
from the first half of the sample and that obtained from the second
differ markedly; the first average is too low and the second average

is too high.  Typically the hydrologist faces a situation intermediate between the two series.  Matalas and Langbein (1962) discuss
the case where the time series for successive flows is

$$x_{i+1} = 0.2\ x_i + \varepsilon_{i+1} \tag{9}$$

where their persistence level of 0.2 is appropriate for many streams.
Matalas and Langbein show that, for that persistence level, the estimate of the mean obtained from 50 successive annual flows is only
as reliable as the estimate obtained from about 33 (independent)
values of a truly random series.  The significance attached to 33
years of independent record is acceptably high, and we must not
overlook the fact that synthetic flows do not improve poor records
but merely improve the quality of designs made with whatever records
are available.  We are faced with the responsibility of using the
available data efficiently.  The chance remains that any finite
statistical sample will be unrepresentative; operational hydrology
operates on the basis of sample statistics alone and certainly cannot change an unrepresentative sample into a representative one.
The engineer using synthetic streamflows should be aware that the
problem of sample bias may exist and also that the probability of
biased results will decrease as the historical sample size increases.  If he has serious suspicions about a badly biased sample,
he may want to apply the regional estimation technique (Benson and
Matalas, 1967) discussed elsewhere in this monograph.

     The second important characteristic of the historical record is
its variability or spread, as measured by its variance or standard
deviation.  The definition of the variance of a population is the
expected value of the square of the difference between a value drawn
at random from the population and the population mean.  Thus if $\mu$ is
the population mean, $x$ is the random variable from the population,
and $E$ is the expectation operator, the variance $\sigma^2$ is defined by

$$\sigma^2 = E[(x - \mu)^2] \tag{10}$$

The standard deviation is the positive square root of the variance,
in this case $\sigma$.  When the analyst has a sample $x_1$, $x_2$, ..., $x_n$ of

values from the population, he uses the following sample estimate of the population variance

$$s^2 = \frac{1}{n-1} \sum_{i=1}^{n} (x_i - \overline{x})^2 = \frac{1}{n-1} \sum_{i=1}^{n} x_i^2 - \frac{1}{n-1} (\overline{x})^2 \qquad (11)$$

where $\overline{x}$ is the sample mean as defined above. The appearance of $n - 1$ in the denominator appears because this computation uses $\overline{x}$ in place of the population mean $\mu$. $S$ is taken as the estimate of $\sigma$.

Figures 8a and 8b show samples with the same mean but with very different standard deviations. As in the case of the mean, it requires a larger sample from a population with persistence to reach a given level of reliability in estimates of the standard deviation than would be sufficient from an independent random population.

The elements of a time series are easily transformed into another series with specified mean and standard deviation. If $t_i$ is a random variable with mean 0 and variance 1, the transformed random variable defined by

$$w_i = \sigma t_i + \mu \qquad (12)$$

will be distributed generally like $t$ but with mean $\mu$ and variance $\sigma^2$ because the linear transformation simply adds a constant and scales the original variable.

Early studies (Hazen, 1914; Sudler, 1927; Barnes, 1954) considered the use of a flow generating scheme that reproduced the historical mean and standard deviation but did not involve any persistence effect. When historical annual flows were considered as a collection of values, it was not inconsistent to assume they derived from a normally distributed population. Consequently, normal random deviates were used to synthesize flow events. In addition, the early work assumed that successive flows were independent of one another. In his studies of long-term storage, Hurst (1951, 1956) drew attention to the existence of persistence in hydrologic sequences. He related dam sizes to the range of the cumulative departures of flows from the mean flow. If $x_1, \ldots, x_n$ are recorded flow values and $\overline{x}$

is their sample mean, then the quantities $y_1, \ldots, y_n$ defined by

$$y_i = x_i - \overline{x} \tag{13}$$

are the departures of the $x_i$ from their mean. The partial sums

$$S_1 = y_1$$
$$S_2 = y_1 + y_2$$
$$\vdots$$
$$S_i = y_1 + y_2 + \ldots + y_i \tag{14}$$
$$\vdots$$

are successive sums of cumulative departures. $S_i$ is the total amount by which the sum of the first $i$ $x$-values exceeds or falls short of $i\overline{x}$, the sum of $i$ mean flows. If $S_{max}$ is the largest of the $S_i$ and $S_{min}$ is the smallest, then $S_{max} - S_{min} = R(n)$ is defined to be the range. Hurst (1951) and independently Feller (1951) found that for independent normal variables the expected value of the range for large $n$ is

$$E[R(n)] = \sigma\left(\frac{\pi}{2}n\right)^{1/2} \approx 1.25\sigma\, n^{1/2} \tag{15}$$

and Feller derived the variance of the range as

$$V[R(n)] = \sigma^2\,(0.272)^2\,N \tag{16}$$

where $\sigma$ is the standard deviation of the flows. Thus the results state that the quantity $E[R(n)]/\sigma$ grows as the square root of the record length. Hurst et al.(1965) considered a large amount of hydrologic and meteorologic data: historical records of streamflows, mud varves, tree rings, and precipitation. They found that the empirical data dictate a faster rate of growth of the range with respect to record length than that predicted by theory. When Chow (1951) applied regression analysis to the Hurst data and fit an exponential equation for the expectation of the range, he found the exponent to be 0.87, which is not consistent with the assumption of independence. This result suggests that the exponent may range

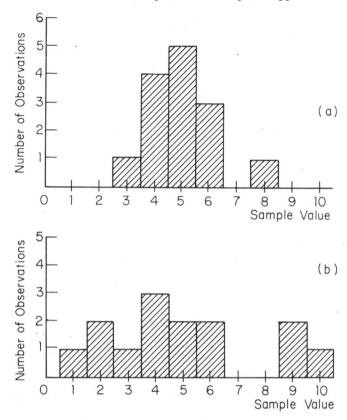

Fig. 8.  (a) Sample with mean = 5; standard deviation = $(20/13)^{\frac{1}{2}}$ = 1.24.  (b)
Sample with mean = 5; standard deviation
= $(100/13)^{\frac{1}{2}}$ = 2.77.

upward from 0.5 depending on the amount of memory (i.e., the extent
of serial correlation) built into the model.  Fiering (1967) has
studied a variety of models and measured their effect on the expected range.

Given $R$ and $S$, the sample standard deviation of inflows to a reservoir, for a sequence of length $N$, Hurst showed that $R/S$ varies as
$N^h$.  Mandelbrot and Wallis (1969a, b, c) have pointed out that for
processes belonging to the Brownian domain of attraction, $h$ tends to
the value 0.5 as $N$ tends to infinity.  Independent processes as well
as autoregressive processes belong to this domain of attraction.
Fractional noise processes, which are outside this domain of attraction, are characterized by exceedingly long memories with parameter $h$,

where $h$ may assume values different from 0.5 and where $h$ does not vary with $N$. Mandelbrot and Wallis have explored the characteristics of fractional noise processes in relation to the generation of synthetic sequences. Statistical properties of estimates of $h$ obtained from independent, lag-one autoregressive, and fractional noise processes have been investigated by Wallis and Matalas (1971). Although the following discussions are confined to processes belonging to the Brownian domain of attraction, one should not assume that these are the only processes for modeling streamflow sequences. As research progresses on fractional noise processes, and perhaps still other processes, their operational capabilities will need to be assessed.

Statistics of the Distribution: Correlation<br>Coefficients and Coefficient of Skewness

In studies that must consider flow persistence, typically associated with wet or dry spells, more elaborate models are needed. The next sample statistic of the historical flows that can be incorporated profitably into the model is the lag-one serial correlation coefficient. The definition of this quantity for the underlying (actual) flow generation model (i.e., the population value of the lag-one serial correlation coefficient) is

$$\rho_1 = \{E[(x_i - \mu)(x_{i+1} - \mu)]\}/\sigma^2 \tag{17}$$

where, as before, $\mu$ is the population mean and $\sigma^2$ is the population variance of flows $x_i$. $\rho_1$ is a measure of the extent to which a flow tends to determine its successor. It is analogous to the product-moment correlation coefficient between two variables in that it measures the closeness with which the variate values may be fit by a straight line. If there is marked persistence in a sequence of flows, there is a notable tendency for both $x_i$ and $x_{i+1}$ to be greater than $\mu$ or for both to be less than $\mu$. Thus there is a distinct tendency for the product $(x_i - \mu)(x_{i+1} - \mu)$ to be positive since it is frequently the product of two terms with the same sign. The expected value of the product is then positive. Conversely, if a higher than average flow is most likely to be followed by a lower than average flow, then the products $(x_i - \mu)(x_{i+1} - \mu)$ tend to be negative, as does their expected value. If, however, there is no persistence in the flow pattern, then a higher than average flow is just as likely to be followed by another high flow as it is to be followed by a lower than average flow. Similarly, a low flow is followed by a high flow or another low flow with equal probability. The $(x_i - \mu)(x_{i+1} - \mu)$ terms are positive as often as they are negative, and their expected value is zero. The variance $\sigma^2$ that appears in the denominator of the expression for $\rho_1$ is a normalizing factor. It restricts the correlation values to the range $[-1, 1]$ and means that correlations from populations with different amounts of spread can be compared meaningfully.

With a finite sample of values $x_1, \ldots, x_n$ drawn from the population, we form the following sample estimate of $\rho_1$:

$$r_1 = \frac{\sum\limits_{i=1}^{n-1} x_i x_{i+1} - \frac{1}{n-1}\left(\sum\limits_{i=1}^{n-1} x_i\right)\left(\sum\limits_{i=2}^{n} x_i\right)}{\left[\sum\limits_{i=1}^{n-1} x_i^2 - \frac{1}{n-1}\left(\sum\limits_{i=1}^{n-1} x_i\right)^2\right]^{0.5}\left[\sum\limits_{i=2}^{n} x_i^2 - \frac{1}{n-1}\left(\sum\limits_{i=2}^{n} x_i\right)^2\right]^{0.5}}$$

$$(18)$$

Higher order correlation coefficients can also be considered. It is reasonable to expect that the flow this year will depend on the flow last year (and the various meteorologic conditions which it reflects) and also on the flow in the next to last year and perhaps on flows in years before that. The population value of the lag-two serial correlation coefficient is defined as

$$\rho_2 = \frac{E[(x_i - \mu)(x_{i+2} - \mu)]}{\sigma^2} \tag{19}$$

Its sample estimate is

$$r_2 = \frac{\sum\limits_{i=1}^{n-2} x_i x_{i+2} - \frac{1}{n-2}\left(\sum\limits_{i=1}^{n-2} x_i\right)\left(\sum\limits_{i=3}^{n} x_i\right)}{\left[\sum\limits_{i=1}^{n-2} x_i^2 - \frac{1}{n-2}\left(\sum\limits_{i=1}^{n-2} x_i\right)^2\right]^{0.5}\left[\sum\limits_{i=3}^{n} x_i^2 - \frac{1}{n-2}\left(\sum\limits_{i=3}^{n} x_i\right)^2\right]^{0.5}}$$

(20)

Similarly, the lag-$k$ serial correlation coefficient is defined by

$$\rho_k = \frac{E[(x_i - \mu)(x_{i+k} - \mu)]}{\sigma^2}$$

(21)

Its sample estimate is

$$r_k = \frac{\sum\limits_{i=1}^{n-k} x_i x_{i+k} - \frac{1}{n-k}\left(\sum\limits_{i=1}^{n-k} x_i\right)\left(\sum\limits_{i=k+1}^{n} x_i\right)}{\left[\sum\limits_{i=1}^{n-k} x_i^2 - \frac{1}{n-k}\left(\sum\limits_{i=1}^{n-k} x_i\right)^2\right]^{0.5}\left[\sum\limits_{i=k+1}^{n} x_i^2 - \frac{1}{n-k}\left(\sum\limits_{i=k+1}^{n} x_i\right)^2\right]^{0.5}}$$

(22)

At this point we must decide how many lags to include in the generating model. One easy but not very restrictive rule is that very long lags should not be used, where very long means near the sample size $n$. For $k$ close to $n$ there are not many pairs of flows separated by $k$ time units and the sample estimate of $\rho_k$ will be unstable and hence very imprecise. Beyond that practical limit it is suggested that if more than one lag is included then additional lags should be included as long as it is practical and profitable to do so. Thus additional lags should be included as long as they produce a model that explains more about the pattern of flows than one with fewer lags does. This decision is discussed more fully in the section on multilag models.

A final statistic of the historical flow sequence which is of particular interest is the coefficient of skewness, defined for the population as

$$\gamma_x = \frac{E[(x - \mu)^3]}{\sigma^3} \tag{23}$$

where $\sigma$ is the population standard deviation. $E[(x - \mu)^3]$ is the third moment about the mean. $\sigma^3$ in the denominator is a scaling factor that renders the statistic dimensionless and thus allows meaningful comparisons of the skewness coefficients of populations with different degrees of spread. The sample estimate of $\gamma_x$ is defined for a sample $x_1, \ldots, x_n$ as

$$\hat{\gamma}_x = \frac{\frac{1}{n} \sum_{i=1}^{n} (x_i - \bar{x})^3}{\left[ \frac{1}{n} \sum_{i=1}^{n} (x_i - \bar{x})^2 \right]^{1.5}} \tag{24}$$

This formula and certain other formulas used in this monograph are slightly biased statistically, although they are none the less widely used. For a discussion of unbiased alternative formulas see Matalas (1958). A computationally useful form of the numerator in this definition is

$$\frac{1}{n} \sum_{i=1}^{n} x_i^3 - \frac{3}{n^2} \left( \sum_{i=1}^{n} x_i^2 \right) \left( \sum_{i=1}^{n} x_i \right) + \frac{2}{n^3} \left( \sum_{i=1}^{n} x_i \right)^3 \tag{25}$$

and the denominator is

$$\left[ \frac{1}{n} \sum_{i=1}^{n} x_i^2 - \bar{x}^2 \right]^{1.5} \tag{26}$$

The coefficient of skewness measures the degree of symmetry (or really the lack of it) of the distribution about its mean. The

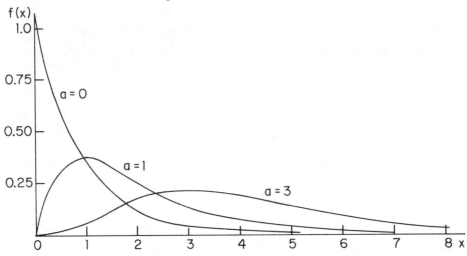

Fig. 9.  Frequency curves of the gamma distribution.

population mean is the center of gravity of the population in the sense that it is the number $m$ that minimizes $E(x - m)^2$ over the population.  If the skewness coefficient is positive, then the part of the population with values greater than the mean is spread farther from the mean than is the part with values less than the mean; that is, the distribution has more weight in its right tail than in its left.  The normal distribution has $\gamma_x = 0$, as does any symmetric distribution.  The family of gamma distributions (a few members are shown in Figure 9) has nonzero skew coefficients.  This family is used in streamflow modeling when the generated flows should produce significant skewness evidenced in the record.

The engineer who is planning a simulation study using synthetic streamflows should start by calculating the mean, standard deviation, lag-one serial correlation coefficient, and coefficient of skewness of the historical flow sequence.  As explained in the next section, the engineer will also need in many cases the mean, standard deviation, lag-one serial correlation coefficient, and coefficient of skewness of the logarithms of the historical flows.

# 3 EVALUATION OF THE STATISTICS

## Selecting a Distribution

The next step is to select the family of distributions for use in the flow generation scheme. A typical history of flows will be quite short, between 10 and 50 years, and consequently the rigorous statistical tests available for testing the goodness of fit of theoretical distributions to large quantities of empirical data cannot aid in the choice of a distribution. The selection of a distribution must involve some intuition and common sense. The distributions generally used are the normal, log normal, and gamma families.

The first major distribution is the normal distribution function (Figure 6b) which is used extensively in statistical applications. The basic justification for the normal function is the central limit theorem which states that a variable which is the sum of identically distributed random variables derived from any distribution with a finite mean and variance is distributed approximately normally. Thus if the flow of a stream in a given time period is the cumulation of different additive factors, then the flows can be considered as the sums of random variables and will be approximately normal.

One helpful tool in testing a historical flow record for normality is normal arithmetic probability paper. With such paper one plots the (sample) values of a variable against the percentage of sample values that are greater than that value and then fits a curve to the points. For a normally distributed random variable, the fitted curve is a straight line (within the limits of sampling errors). The sample coefficient of skewness is another aid in testing normality; the skewness coefficient of the normal distribution is zero, and the sample skewness coefficient of a sample from a normal distribution should be close to zero. The population of flows can never be

33

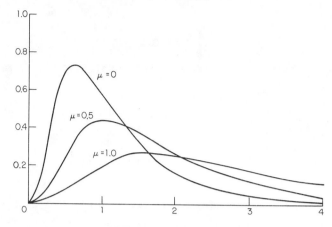

Fig. 10.   Frequency curves of the log
normal distribution (μ is the mean).

exactly normal, we note, since any normal distribution assigns some
nonzero probability to negative values of the random variable and
flows are certainly not negative.  This limitation does not, however,
preclude the use of the normal distribution.  It is most likely that
none of the distributions tried will fit exactly, anyway, and if the
normal fits better than the others, we can use the normal and simply
discard any negative flows that are generated.

The second major distribution widely used in streamflow modeling
is the log normal distribution as shown in Figure 10 (Aitchison and
Brown, 1957).  If $y$ is a random variable with the property that
log $y$, the logarithm of $y$ to any base, is normally distributed, then
$y$ has a log normal distribution.  The log normal distribution has
the convenient property that it is undefined for negative values so
that log normal generation will not produce negative flows.  It also
provides more resolution in the range of low flows than does the
normal distribution and hence may be particularly useful for studies
of droughts.  To test whether a sample might come from a log normal
distribution, one takes the logarithms of the flows and then applies
the normality tests described above to the logs; alternatively, log
probability paper is available for use as detailed above, and its
use saves computation time.

The third major distribution used in flow generation models is
the gamma distribution; a few members of the gamma family are

plotted in Figure 9. The gamma distribution is used when the historical record shows distinct skewness of the flows and of their
logarithms and when skewness should be demonstrated by the synthetic
flow sequences. As noted further in the section on multilag models,
the gamma distribution cannot be used in models that involve more
than a single lag because no general theory is available.

In trying to select an appropriate distribution, the analyst
starts by plotting sample values and considering sample moments.
Unfortunately, in many cases these procedures will not provide an
easy choice, for it is quite likely that no one of the distributions
will fit significantly better than the others. The analyst must
then consider the purposes of the proposed simulation and must as-
sess the expected costs of making incorrect decisions. As mentioned
above, the log normal distribution provides especially good resolution in the range of low flows because small changes in low numbers
produce relatively large changes in the logarithms of those numbers.
With a log normal model, the planner assumes that the logs of flows
are normally distributed. The gamma distribution is the only one of
the three considered here that shows skewness of both flows and
their logs; therefore it is appropriate for extensions of historical
flows in which skewness is marked and important. The hydrologist
considers the physical characteristics of the particular basin under
study in making this decision. If past history can be considered to
influence this year's flow only through its effects in determining
flows in the last one or two years, then the gamma family might be
appropriate. If the effect of the past is significantly more complicated and the hydrologist feels that multiple lags are important,
then the gamma distribution must be excluded.

If no clear decision is indicated by the previous analysis, the
planner should invoke economic considerations in his further selection. For example, if he cannot choose between two possible distributions for flows, the planner can conduct two simulation studies,
one for each of the possible distributions. He can then invoke decision and game theories in choosing one of the two distributions

and its associated optimal design.  We return to this question at
the end of the monograph.

This discussion is not meant to advocate any particular distribution.  It is meant rather to emphasize the importance of well considered, economically realistic criteria for decision making.  No
matter how he chooses to apply operational hydrology in planning, the
engineer's decision will have important economic implications.  If he
chooses not to simulate but to decide without such evaluation he is
still making a decision that will have economic consequences.  Our
point is that it is far better for the planner to acknowledge his
economic benefit and loss functions even if it is difficult to specify them.  Failure to do so simply means that he is basing his
choices on an evaluation scheme that is inarticulated, is often ill-
considered, and in many cases is perhaps even at odds with the pri-
orities the planner would express if he were forced to voice his
views.  The carefully selected benefit functions should be used both
to choose the optimal design for a given distribution of natural
flows and to choose among possible flow distributions when several
seem reasonable and acceptable.

There is one fact that should comfort the engineer who is still
uneasy about selecting a distribution.  Empirical studies have shown
that the mean and standard deviation are much more important than
other statistics in producing good results in most basin simulation
studies.  In many cases it is sufficient to reproduce the historical
mean and standard deviation; therefore the normal distribution is
quite adequate.  Fortunately the mean and standard deviation can be
estimated tolerably well from moderate sized samples.  Usually the
estimates obtained from 30 to 50 years of historical records are
very good.  Other statistics, such as higher order moments (the
skewness coefficient, for example) and correlation coefficients for
large lags, are subject to more pronounced sampling errors; therefore analysts may hesitate to use sample estimates.  Fortunately
these less stable statistics do not seem crucial for rational evaluation of alternative schemes.

### Refinements in Evaluating Sample Statistics

Sometimes the hydrologist feels strongly that for some reason a particular flow sequence is simply not representative; for example, a very short record might cover an unusually dry period in the region of the stream. At other times, a hydrologist might want to model a stream for which the historical record is extremely short or even nonexistent. Benson and Matalas (1967) suggest a procedure for such circumstances. They advocate using all the long streamflow and related meteorologic records from a region to improve or originate estimates of the parameters of flow in a given stream. To do this, they use multiple regression analysis to express the means, for example, as functions of the physical characteristics and climatic variables for the basins of the respective streams. Next they substitute the physical characteristics and climatic variables for the given stream into a regression equation to estimate the mean flow for that stream. A second regression estimates the standard deviation of the flows as a function of the physical characteristics and climatic variables. Similar procedures give estimates of correlation and skewness coefficients.

Benson and Matalas use their procedure for the Potomac River basin to derive regression equations of the form

$$Y = aA^{b_1} S^{b_2} St^{b_3} p^{b_4} Sn^{b_5} F^{b_6} \tag{27}$$

where

> $Y$, statistical parameter to be estimated;
> $A$, drainage area, square miles;
> $S$, slope of the main channel, feet per mile;
> $St$, surface storage area in lakes and ponds, percent of the total drainage area (increased by 1%);
> $Sn$, annual snowfall, inches;
> $p$, mean annual precipitation, inches;
> $F$, forested area, percent of total area.

The parameters $a$, $b_1$, ..., $b_6$ are regression coefficients determined by the regression analysis. Benson and Matalas report finding

satisfactory equations for the means and standard deviations but
could not find good equations for the correlation and skewness coefficients.  They suggest using graphical methods and averages to estimate these parameters.

Finally, we should consider time intervals for the simulation,
whether to generate flows for years, months, or some other time interval.  The purpose of the proposed design is, of course, a basic
determining factor.  If the main purpose of a new dam is to provide
seasonal storage so that flow from the wet season can be saved for
use during the dry season, then annual flows are clearly not of much
use in evaluating alternate designs, and flows for shorter periods
are needed.  In fact, the planner may want to use different time intervals for different parts of the year.  Once the wet season starts
he may be virtually certain that there will be enough water for use
during that season and he may care only about total wet season flow.
During the dry season, on the other hand, he may want to monitor the
storage level much more closely because of the importance of the exact patterns of flows and drafts.  Consequently, monthly, weekly, or
even daily flows may be needed during the dry season.  If a dam is
meant to provide seasonal storage and overyear storage to accommodate droughts or floods, then both seasonal and annual flows are
important.

To generate such flows the analyst needs estimates of the parameters of flows for the relevant time periods.  Sometimes historical
data are very hard or impossible to find.  This lack of data may
even preclude the use of particularly interesting time intervals or
the use of such intervals with flows that follow a historical distribution.  Annual and monthly flows, or flows for other intervals,
should be examined separately and tested for appropriate distribution.  Annual and monthly flows may well come from different families of distributions.  Harms and Campbell (1967) use normal annual
flows and log normal monthly flows in studies of Washington rivers,
for example.  When flows for different time intervals are used,
problems arise in making these flows consistent; for example, the

problems of making monthly flows add up to annual flows or of making month to month and year to year correlations of flows realistic. These problems have not been solved completely. Harms and Campbell suggest one method of producing consistency with respect to most of the correlations when a model incorporating annual and monthly flows is used. We discuss their method of consistency after describing several generating schemes.

Fiering (unpublished memorandum, 1970) considers a 2-season (wet and dry) model, with $\beta_{21}$ being the regression of dry season flow (2) on wet season flow (1) within the same calendar year and $\beta_{12}$ being the regression of wet season flow (1) on the dry season flow of the previous year. If $\rho_{21}$ and $\rho_{12}$ are defined analogously, and if $\rho$ is the lag-one annual serial coefficient, then the theory of Markov processes shows that

$$\rho = \frac{\rho_{12}(\sigma_1 + \rho_{21}\sigma_2)(\sigma_2 + \rho_{21}\sigma_1)}{\sigma_1^2 + \sigma_2^2 + 2\rho_{21}\sigma_1\sigma_2} \tag{28}$$

where $\sigma_1$ and $\sigma_2$ are the standard deviations of flows in seasons 1 and 2, respectively. Thus, if the overyear seasonal correlation $\rho_{12} = 0$, then $\rho = 0$. However, even if the with-in-year seasonal correlation $\rho_{21} = 0$, then $\rho$, the annual correlation, is not necessarily zero. Of course, more elaborate expressions can be derived for years with more than two seasons.

In many cases the observed annual correlation, denoted by $\rho^*$, is widely different from the theoretical value of $\rho$, which suggests that some or all of the generating processes are nonMarkovian. Thus all the seasonal and annual correlations cannot be maintained simultaneously, and some compromises must be struck so that a consistent set of estimators of the $\rho_{ij}$ and $\rho$ values can be found. The exact choice depends on the criterion for calculating the regression coefficients.

### Summary of Calculations for Estimating Parameters and Selecting a Distribution

1. Examine the historical record. If man's activities noticeably altered a particular basin some years ago, then earlier flows for a stream in the basin may not be relevant for estimating streamflow parameters. If flows in a stream have been regulated in the past, the analyst will want to adjust the record to find the flows that would have occurred had there been no regulation. If he cannot follow this plan, he may try to divide flow at a given point into regulated and unregulated flow. The unregulated flows can be synthesized as described here; the regulated flows (perhaps just taken as equal to their target flows) can then be added to unregulated flow to reconstitute total flows.

Thus, select a set of relevant observed or estimated flows $x_1, \ldots, x_n$. If no historical flows are available or if there is reason to believe that the available historical sequence is unrepresentative, consider using the regional estimation technique described in the previous section.

2. Calculate the mean $\bar{x}$ of the historical flows for whatever time intervals are considered important. (If the planned simulation involves different kinds of intervals, such as months and years, follow this step and all other steps for flows for all relevant intervals; calculate the mean of the monthly flows and the mean of the annual flows, for example.) The mean $\bar{x}$ is defined by

$$\bar{x} = \frac{1}{n} \sum_{i=1}^{n} x_i \tag{29}$$

It is the average flow.

3. Calculate the standard deviation $s$ of the historical flows, by

$$s^2 = \frac{1}{n-1} \sum_{i=1}^{n} x_i^2 - \frac{n}{n-1} \bar{x}^2 \tag{30}$$

4.  Calculate the skewness coefficient $\hat{\gamma}_x$ for the sample. It is calculated by

$$\hat{\gamma}_x = \frac{\displaystyle\sum_{i=1}^{n} x_i^3 - 3\bar{x}\sum_{i=1}^{n} x_i^2 + 2n\bar{x}^3}{n\left(\dfrac{1}{n}\displaystyle\sum_{i=1}^{n} x_i^2 - \bar{x}^2\right)^{1.5}} \tag{31}$$

5.  Calculate $y_1, y_2, \ldots, y_n$ where $y_i$ is the logarithm of the flow value $x_i$. Calculate the mean, standard deviation, and skewness coefficient of the $y$ values.

6.  Use normal probability paper to check whether the flows can be considered to be drawn from a normal population. Let $z_i, \ldots, z_n$ be a rearrangement of $x_1, \ldots, x_n$ in increasing size order. For each $z_i$ calculate the fraction of the sample with values at least as large as $z_i$; this fraction is $(n - i + 1)/n$. Plot $z_i$ versus $(n - i + 1)/n$ on the probability paper and fair in a curve. If the best fitting curve is straight or nearly so, the assumption of normality is appropriate. A further check is that $\hat{\gamma}_x$ should be close to zero if the flows are normally distributed.

7.  If step 6 suggests that flows are not clearly normal, repeat step 6 for the $y$ values (as calculated in step 5). If the $y$ seem normal, an assumption of log normality is appropriate. Use the flow generating technique developed by Matalas (1967) and described in the following sections, and generate new $y$ values (not $x$ values directly). In the generation techniques for flows, it is not possible to reproduce the historical statistical parameters of both the flows and their logarithms. Since we are primarily interested in the flows, we choose to maintain the flow parameters and, as described in the following sections, we calculate parameters for use in logarithm generating schemes to do so. The result of such a generation scheme is a sequence $y_1^*, y_2^*, \ldots$ of logarithms of flows. Then $\exp(y_1^*), \exp(y_2^*), \ldots$ is a sequence of synthetic flows.

8.  If neither the assumption of normality nor the assumption of log normality is clearly indicated by steps 6 and 7, several

alternatives are possible.  If the skewness is pronounced ($\hat{\gamma}_x$ is significantly different from zero), if the modeler wants the synthetic flows to reflect such skewness, and if no more than two lags seem necessary in the model, the gamma distribution is appropriate. Otherwise, consider the decision theoretic techniques described in the section on game and decision theories for deciding which distribution is appropriate for the flows.

9.  Calculate the first serial correlation coefficient $r_1$ (or lag one serial correlation coefficient).  (Note that if the log normal distribution has been selected, the references to flows in this and later calculations should be interpreted as references to logs of flows since adopting the log normal assumption implies generating logs of flows.)  The correlation $r_1$ can be calculated as

$$r_1 = \frac{\displaystyle\sum_{i=1}^{n-1} x_i x_{i+1} - \frac{1}{n-1}\left(\sum_{i=1}^{n-1} x_i\right)\left(\sum_{i=2}^{n} x_i\right)}{\left[\displaystyle\sum_{i=1}^{n-1} x_i^2 - \frac{1}{n-1}\left(\sum_{i=1}^{n-1} x_i\right)^2\right]^{0.5}\left[\displaystyle\sum_{i=2}^{n} x_i^2 - \frac{1}{n-1}\left(\sum_{i=2}^{n} x_i\right)^2\right]^{0.5}}$$

(32)

For multilag studies, the correlation coefficient for higher order lags $r_k$ can be calculated by

$$r_k = \frac{\displaystyle\sum_{i=1}^{n-k} x_i x_{i+k} - \frac{1}{n-k}\left(\sum_{i=1}^{n-k} x_i\right)\left(\sum_{i=k+1}^{n} x_i\right)}{\left[\displaystyle\sum_{i=1}^{n-k} x_i^2 - \frac{1}{n-k}\left(\sum_{i=1}^{n-k} x_i\right)^2\right]^{0.5}\left[\displaystyle\sum_{i=k+1}^{n} x_i^2 - \frac{1}{n-k}\left(\sum_{i=k+1}^{n} x_i\right)^2\right]^{0.5}}$$

(33)

## Example of Parameter Calculation

Table 4 shows the annual flows for the Selway River in Idaho. The data will be used to demonstrate the computations which are explained in the preceding pages and which are necessary for streamflow simulation.

Mean flow (equation 7): $\bar{x} = \sum\limits_{i=1}^{20} x_i/20 = 11776/20 = 588.8$ haf

Variance (equation 11): $s^2 = \dfrac{1}{20-1}\left(\sum\limits_{i=1}^{20} x_i^2 - 20\bar{x}^2\right)$

$$= 1/19 \ (7,500,172 - 20(588.8)(588.8)$$

$$= 29,813.85 \ (haf)^2$$

Standard deviation: $s = \sqrt{s^2} = \sqrt{29.813.85} = 172.667$ haf

Skewness (equations 25-26): $g$

$d_1 = \sum\limits_{i=1}^{20} x_i^3/20 = 5,177,696,020/20$

$$= 258,884,801 \ (haf)^3$$

$d_2 = 3\left(\sum\limits_{i=1}^{20} x_i^2\right)\left(\sum\limits_{i=1}^{20} x_i\right)/20^2 = 3(7,500,172)(11,776)/400$

$$= 662,415,191 \ (haf)^3$$

$d_3 = 2\left(\sum\limits_{i=1}^{20} x_i\right)^3/20^3 = 11,776^3/4000$

$$= 408,256,774 \ (haf)^3$$

$d_4 = \left(\dfrac{1}{20}\sum\limits_{i=1}^{20} x_i^2 - \bar{x}^2\right)^{0.5} = (7,500,172/20) - (588.8)(588.8)$

$$= 168.295 \ haf$$

$$g = \frac{d_1 - d_2 + d_3}{d_4^3} = 0.9916$$

*Synthetic Streamflows*

TABLE 4.  Data from Annual Flows, Selway

| $i$ | $x_i$ | $x_i{}^2$ | $x_i{}^3$ | $x_i x_{i+1}$ | $x_i x_{i+2}$ | $x_i - \bar{x}$ |
|---|---|---|---|---|---|---|
| 1 | 429 | 184041 | 78953589 | 161304 | 220935 | -159.8 |
| 2 | 376 | 141376 | 53157376 | 193640 | 135360 | -212.8 |
| 3 | 515 | 265225 | 136590875 | 185400 | 300760 | -73.8 |
| 4 | 360 | 129600 | 46656000 | 210240 | 162360 | -228.8 |
| 5 | 584 | 341056 | 199176704 | 263384 | 254040 | -4.8 |
| 6 | 451 | 203401 | 91733851 | 196185 | 217833 | -137.8 |
| 7 | 435 | 189225 | 82312875 | 210105 | 321030 | -153.8 |
| 8 | 483 | 233289 | 112678587 | 356454 | 393645 | -105.8 |
| 9 | 738 | 544644 | 401947272 | 601470 | 329886 | 149.2 |
| 10 | 815 | 664225 | 541343375 | 364305 | 463735 | 226.2 |
| 11 | 447 | 199809 | 89314623 | 254343 | 245850 | -141.8 |
| 12 | 569 | 323761 | 184220009 | 312950 | 400576 | -19.8 |
| 13 | 550 | 302500 | 166375000 | 387200 | 585750 | -38.8 |
| 14 | 704 | 495616 | 348913664 | 749760 | 481536 | 115.2 |
| 15 | 1065 | 1134225 | 1207949625 | 728460 | 814725 | 476.2 |
| 16 | 684 | 467856 | 320013504 | 523260 | 416556 | 95.2 |
| 17 | 765 | 585225 | 447697125 | 465885 | 505665 | 176.2 |
| 18 | 609 | 370881 | 225866529 | 402549 | 326424 | 20.2 |
| 19 | 661 | 436921 | 288804781 | 354296 | | 72.2 |
| 20 | 536 | 287296 | 153990656 | | | -52.8 |
| $\Sigma$ | 11776 | 7500172 | 5177696020 | 6921190 | 6576666 | 0.0 |

$w_i$ is defined as $(x_i - \bar{x})/s$ where $s$ is the standard deviation of the flows. ($\Sigma w_i$ is 19.0 rather than 20.0 because the sample value of $s = 172.67$ was used to calculate $w_i$.)

Lag-one serial correlation coefficient (equation 32): $r_1$

$$f_1 = \sum_{i=1}^{19} x_i\, x_{i+1} = 6,921,190\ (haf)^2$$

$$f_2 = \sum_{i=1}^{19} x_i = 11,240\ haf$$

$$f_3 = \sum_{i=1}^{20} x_i = 11,347\ haf$$

River, Idaho in Hundreds of Acre-Feet

| $\sum_{i=1}^{k} (x_i - \bar{x})$ | $w_i$ | $w_i^2$ | $\log_e x_i$ | $(\log_e x_i)^2$ |
|---|---|---|---|---|
| -159.8 | -0.9255 | 0.8565 | 6.0615 | 36.7413 |
| -372.6 | -1.2324 | 1.5189 | 5.9296 | 35.1600 |
| -446.4 | -0.4272 | 0.1827 | 6.2442 | 38.9896 |
| -675.2 | -1.3251 | 1.7559 | 5.8861 | 34.6462 |
| -680.0 | -0.0278 | 0.0008 | 6.3699 | 40.5756 |
| -817.8 | -0.7981 | 0.6369 | 6.1115 | 37.3500 |
| -971.6 | -0.8907 | 0.7934 | 6.0753 | 36.9098 |
| -1077.4 | -0.6127 | 0.3755 | 6.1800 | 38.1926 |
| -928.2 | 0.8641 | 0.7467 | 6.6039 | 43.6121 |
| -702.0 | 1.3100 | 1.7162 | 6.7032 | 44.9327 |
| -843.8 | -0.8212 | 0.6744 | 6.1026 | 37.2417 |
| -863.6 | -0.1147 | 0.0131 | 6.3439 | 40.2448 |
| -902.4 | -0.2247 | 0.0505 | 6.3099 | 39.8151 |
| -787.2 | 0.6672 | 0.4451 | 6.5568 | 42.9913 |
| -311.0 | 2.7579 | 7.6061 | 6.9707 | 48.5911 |
| -215.8 | 0.5514 | 0.3040 | 6.5280 | 42.6142 |
| -39.6 | 1.0205 | 1.0413 | 6.6399 | 44.0880 |
| -19.4 | 0.1170 | 0.0137 | 6.4118 | 41.1114 |
| 52.8 | 0.4181 | 0.1748 | 6.4938 | 42.1688 |
| 0.0 | -0.3058 | 0.0935 | 6.2841 | 39.4903 |
| | 0.0000 | 19.0000 | 126.8066 | 805.4663 |

$$f_4 = \sum_{i=1}^{19} x_i^2 - \frac{1}{19}\left(\sum_{i=1}^{19} x_i\right)^2 = 7,212,876 - \frac{1}{19}(11,240)^2$$

$$= 563,528.6 \ (haf)^2$$

$$f_5 = \sum_{i=2}^{20} x_i^2 - \frac{1}{19}\left(\sum_{i=2}^{20} x_i\right)^2 = 7,316,131 - \frac{1}{19}(11,347)^2$$

$$= 539,583 \ (haf)^2$$

$$\sqrt{f_4} = 750.685 \ haf$$

$$\sqrt{f_5} = 734.563 \ haf$$

$$r_1 = \frac{f_1 - \frac{1}{19}(f_2 \cdot f_3)}{\sqrt{f_4}\sqrt{f_5}} = \frac{208,543.7}{551,425.9} = 0.3782$$

Note: Sometimes a different formula is suggested for the lag-one serial correlation:

$$r_1{}^* = \frac{\sum\limits_{i=1}^{n-1} x_i\, x_{i+1} - (n-1)\bar{x}^2}{(n-2)s^2}$$

This formula should not be used; it is biased and for the small sample sizes typical of streamflow data the bias can be pronounced. For the data in Table 4, $r_1{}^* = 0.622$, a value significantly different from $r_1 = 0.3782$.

Lag-two serial correlation coefficient (equation 33): $r_2$

$$k_1 = \sum_{i=1}^{18} x_i\, x_{i+2} = 6{,}576{,}666 \text{ (haf)}^2$$

$$k_2 = \sum_{i=1}^{18} x_i = 10{,}579 \text{ haf}$$

$$k_3 = \sum_{i=3}^{20} x_i = 10{,}971 \text{ haf}$$

$$k_4 = \sum_{i=1}^{18} x_i{}^2 - \frac{1}{18}\left(\sum_{i=1}^{18} x_i\right)^2 = 6{,}775{,}955 - \frac{1}{18}(10{,}579)^2$$
$$= 558{,}441.6 \text{ (haf)}^2$$

$$k_5 = \sum_{i=3}^{20} x_i{}^2 - \frac{1}{18}\left(\sum_{i=3}^{20} x_i\right)^2 = 7{,}174{,}755 - \frac{1}{18}(10{,}971)^2$$
$$= 487{,}930.5 \text{ (haf)}^2$$

$$\sqrt{k_4} = 747.290 \text{ haf}$$

$$\sqrt{k_5} = 698.520 \text{ haf}$$

$$r_2 = \frac{k_1 - \frac{1}{18}(k_2 \cdot k_3)}{\sqrt{k_4}\,\sqrt{k_5}} = 0.2467$$

Mean of standardized variates $(w_i\text{'s}) = \bar{w} = \dfrac{1}{20} \sum\limits_{i=1}^{20} w_i = 0$

Variance of standardized variates $= \dfrac{1}{19}\left(\sum\limits_{i=1}^{20} w_i^2 - 20\bar{w}^2\right)$

$$= \dfrac{\Sigma w_i^2}{19} = 1.0$$

Maximal value of $\sum\limits_{i=1}^{k} (x_i - \bar{x}) = 52.8$ haf (see equation 14)

Minimal value of $\sum\limits_{i=1}^{k} (x_i - \bar{x}) = -1077.4$ haf (see equation 14)

Range $= 52.8 - (-1077.4) = 1130.2$ haf

Mean of logarithms: $\bar{z} = \dfrac{1}{20} \sum\limits_{i=1}^{20} \log_e x_i = \dfrac{126.8066}{20} = 6.340$

Natural logarithm of mean $= \log_e (588.8) = 6.378$

Variance of logarithms $= \dfrac{\sum\limits_{i=1}^{20} (\log_e x_i)^2 - 20\bar{z}^2}{19} = 0.077$

Natural logarithm of variance $= 10.303$

Standard deviation of logarithms $= 0.278$

Natural logarithm of standard deviation $= 5.151$

# 4 CHOICE OF A MODEL

## Markov Model

After the sample parameters of the flows are identified and after
a distribution is selected, the modeler must next select a generating scheme for the synthetic flows. As we saw in the section on
time series and random numbers, the form of such a scheme is

$$q_i = d_i + e_i \qquad (34)$$

with $d_i$ the deterministic part and $e_i$ the random part of the synthetic flow. Because flows that are serially independent do not
seem adequate for most hydrologic modeling, we want to include a
nonzero deterministic component to reflect persistence in the generation process. We assume that the $e_i$ in the generation equation are
independently distributed with mean zero and constant variance.
Further, we make the very important assumption that the correlation
between two different flows depends only on the time interval between the flows; we assume that the basin does not change in such a
way that the expected persistence between one flow and the next varies with time. This assumption also means that the degree of persistence between successive flows does not depend on the level of
those flows. (Current research is directed at the problem of estimating β, the regression coefficient, as a function of flow.) In
actual flow patterns there is more persistence between low flows
than between high ones, so this assumption is not completely realistic. However, we make this simplifying assumption to reduce the
problem to manageable complexity. The form of the deterministic
component will then have the linear autoregressive form

$$d_i = \beta_o + \beta_1 q_{t-1} + \beta_2 q_{t-2} + \cdots + \beta_m q_{t-m} \qquad (35)$$

where $d_i$ is a linear combination of $m$ previous flow values, for some finite $m$. The simplest model of this form is

$$q_i = \beta_0 + \beta_1 \, q_{i-1} + e_i \tag{36}$$

where $m = 1$ (e.g., Fiering, 1967). The model assumes that the entire influence of the past on the current flow is reflected in the previous flow value. The model is thus quite restrictive (and certainly not phenomenologically correct), but it is a distinct improvement over the independence model and produces flows that are useful in many planning problems.

In this lag-one or Markovian flow model we specify next the constants $\beta_0$ and $\beta_1$ and the exact form of the $e_i$ or error terms. First we consider normal flows. If $\mu$ is the mean flow value, $\rho$ is the lag-one serial correlation coefficient, and $\sigma^2$ is the variance of the flows, we start with the form

$$q_i = \mu + \rho \, (q_{i-1} - \mu) + e_i \tag{37}$$

That is, we assume that the flow $q_i$ is the sum of the mean, a proportion (given by $\rho$) of the departure of the previous flow from its mean, and a random component $e_i$. So far we assume only that the $e_i$ have zero mean and constant variance and that they do not depend on $q_{i-1}$; if the $q_i$ are to be normally distributed then the $e_i$ must be normal. The $q_i$ as specified by this equation have mean (or expected value) $\mu$, so that the suggested form gives the desired average value. The $q_i$ have variance

$$E\left[\mu + \rho(q_{i-1} - \mu) + e_i\right]^2 - \mu^2 = \rho^2 \sigma^2 + \sigma_e^2 \tag{38}$$

where $\sigma_e^2$ is the variance of the random components $e_i$. The variance of the $q_i$ is related to $\sigma_e^2$ by

$$\sigma_e^2 = \sigma^2 \, (1 - \rho^2) \tag{39}$$

If $t_i$ is a normally distributed, serially independent random variable with zero mean and unit standard deviation, then $t_i \sigma \sqrt{(1 - \rho^2)}$
is a normally distributed, serially independent variable with zero
mean and variance $\sigma^2(1 - \rho^2)$. The generating equation

$$q_i = \mu + \rho(q_{i-1} - \mu) + t_i \sigma \sqrt{(1 - \rho^2)} \qquad (40)$$

where the $t_i$ are independent normal sampling deviates with mean 0 and
standard deviation 1, gives normally distributed synthetic flows that
preserve the mean, variance, and first order correlation coefficient
of the historical flow record.

Because the normal distribution assigns nonzero probability to
negative values, this procedure will occasionally give a negative
flow. Negative flows should be used in the generating equation to
produce succeeding flows and then should be discarded; they should
not be used as flows for simulation.

We should emphasize again the meaning of such a Markovian flow
scheme. It says that a given flow depends on the preceding flow and
a random component and on these factors alone. The notion of dependence on previous flow should be quite easy to accept. One explanation might be that a high flow in one time period will build up
groundwater levels and thus provide a tendency toward high flow in
the next period, even without heavy precipitation in that period.
Similarly, the groundwater supply will be depleted in periods of low
flow; even if rainfall is heavy in the following period, much of the
precipitation enters the soil and the groundwater aquifer and
streamflows will be low. Of course the validity of the Markov model
for planning does not depend on any particular explanation; it merely depends of the existence of persistence. Dependence on exactly
one previous flow is difficult to justify. We consider this assumption more fully in the section on multilag models.

It is a property of a normal Markov process that elements in the
generated series that are $k$ units apart have correlation $\rho^k$; i.e.,
the lag-$k$ serial correlation coefficient in $\rho^k$. Figure 11 shows a
plot of $k$ versus the lag-$k$ correlation coefficient for a Markov

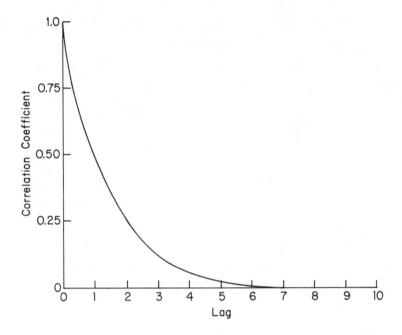

Fig. 11.  Theoretical Markovian correlogram.

process when a curve has been drawn between the correlation values
for integral $k$.  Such a plot is called a correlogram, and the sample
shows that the correlation between elements of the series decays ex-
ponentially with increasing separation between the elements.

The Markov scheme

$$q_i = \mu + \rho(q_{i-1} - \mu) + e_i \tag{41}$$

can also be adapted to give flows that are distributed approximately
as gamma with mean $\mu$, variance $\sigma^2$, and skewness $\gamma_x$ (Thomas and
Fiering, 1963).  First introduce $\gamma_\xi$ as the skewness of the random
component of each flow event.  It is different from the skewness of
the flows because the sums of gamma variates, not nearly as attract-
ive as those of normal variates, are not necessarily gamma.  Next
define

$$\gamma_\xi = \frac{1 - \rho^3}{(1 - \rho^2)^{1.5}} \, \gamma_x \tag{42}$$

where $\rho$ is the lag-one correlation coefficient of the flows and $\gamma_x$ is the skewness coefficient of the flows. Let $t_i$ be normally distributed with zero mean and variance 1. Then a modified random sampling variate $\xi_i$ is defined by (Wilson and Hilferty, 1931)

$$\xi_i = \frac{2}{\gamma_\xi} \left( 1 + \frac{\gamma_\xi t_i}{6} - \frac{\gamma_\xi^2}{36} \right)^3 - \frac{2}{\gamma_\xi} \tag{43}$$

This deviate is distributed approximately as gamma with mean zero, variance 1, and coefficient of skewness $\gamma_\xi$. Moreover, its use preserves the third moment of the recorded flows. If the sequence of $\xi_i$ is used in the Markov scheme

$$q_i = \mu + \rho(q_{i-1} - \mu) + \xi_i \, \sigma \sqrt{(1 - \rho^2)} \tag{44}$$

the resulting sequence of $q_i$ will be distributed approximately as gamma with mean $\mu$, variance $\sigma^2$, first correlation coefficient $\rho$, and skewness $\gamma_x$. This type of transformation is generally applicable for schemes that aim to preserve any number of moments; its use is not recommended even with uncommonly long records because of the extreme instability of higher moments.

Another procedure for flows distributed as gamma was suggested by Yevjevich and is discussed by Matalas (1967). That procedure begins with a long sequence $y_1$, $y_2$, ... of numbers with zero mean, unit variance, and lag-one serial correlation coefficient $\rho_y$, which have been formed with the normal Markov generation scheme. Then an integer $m$ is selected and a new sequence $z_1$, $z_2$,... is formed by

$$z_i = \sum_{j=(i-1)m+1}^{im} y_j^2 \tag{45}$$

Thus if the sequence of $y$ is divided into subsequences of length $m$, $z_i$ is the sum of the squares of the elements in the $i$th such subsequence. The $z$ are distributed as gamma with mean $m$, variance $2m$, skewness coefficient $2\sqrt{2/m}$, and lag-one serial correlation coefficient $\rho_y^2$. Finally, a sequence $q_1$, $q_2$, ... of synthetic flows is formed from

$$q_i = \bar{x} + s(x)[(2m)^{-0.5} z_i - (m/2)^{0.5}] \tag{46}$$

where $\bar{x}$ and $s(x)$ are the mean and standard deviation of the historical flow record. The $q$ maintain the mean, variance, and lag-one serial correlation coefficient of the historical flows; the skewness of the $q$ is $(8/m)^{0.5}$. Since $m$ must be an integer, this procedure cannot generate flows with arbitrary skewness coefficients; in particular, the skewness cannot be greater than $2\sqrt{2}$. If this scheme is appropriate, the planner selects the $m$ for which $(8/m)^{0.5}$ is closest to $q(x)$, which is the sample skewness coefficient for the historical flows.

When the modeler uses the Markov generation scheme to produce logarithms of flow he must remember that the procedure reproduces the mean, variance, serial correlation coefficient, and skewness coefficient of the logs of the flows. The serial correlation and the skewness coefficient of the flows themselves will not necessarily be preserved. In practice, the resultant distortion may be important and Matalas (1967) has suggested procedures for ensuring that the moments of the flows are maintained.

Matalas assumes that the number $a$ is a lower bound on the possible flow values and that if $x$ denotes a flow, then $y = \log(x - a)$ is normally distributed. The parameters of the $x$ are related to the parameters of the $y$ as follows:

$$\mu(x) = a + \exp\left[\sigma^2(y)/2 + \mu(y)\right] \tag{47}$$

$$\sigma^2(x) = \exp\left\{2[\sigma^2(y) + \mu(y)]\right\} - \exp\left[\sigma(y) + 2\mu(y)\right] \tag{48}$$

$$\gamma(x) = \frac{\exp\left[3\sigma^2(y)\right] - 3\exp\left[\sigma^2(y)\right] + 2}{\left\{\exp\left[\sigma^2(y)\right] - 1\right\}^{3/2}} \tag{49}$$

$$\rho_1(x) = \frac{\exp\left[\sigma^2(y)\,\rho_1(y)\right] - 1}{\exp\left[\sigma^2(y)\right] - 1} \tag{50}$$

To preserve the historical statistics of the flows rather than of their logarithms, Matalas suggests calculating the sample statistics $\bar{x}$, $s^2(x)$, $g(x)$, and $r_1(x)$ and substituting these values into the four equations above. The equations can then be solved for values of $\mu(y)$, $\sigma^2(y)$, $\rho_1(y)$, and $a$. These estimates, not the sample statistics of the logarithms of historical flows, are used in the flow generation process to give a series $h_1$, $h_2$, ... of synthetic logarithms of flows. When a series of synthetic flows is formed from

$$q_i = \exp(h_i) + a \tag{51}$$

the flows have expected parameters $\bar{x}$, $s^2(x)$, $g(x)$, and $r_1(x)$, as desired.

Finally, we note that the Markov models discussed above assume that all time periods, taken here to be years, are characterized by identically distributed flows; the flows in each period have mean $\mu$, variance $\sigma^2$, and serial correlation $\rho$. For studies that use only annual flows this assumption seems reasonable and can generally be defended unless long-term trends are at work. However, for studies that require flows for seasons, months, or other subdivisions of the year, more complicated models are necessary. Certainly the mean flow for a month or other period during a wet season is different from the mean flow for the same duration in a dry season. The

Markov model can be modified to reflect different means (and variances) in different seasons with very little additional conceptual complexity; the notation and computational bookkeeping become noticeably more complicated.

First, we consider the normal Markov model with $m$ different seasons (months or other time intervals) each year. The model now uses two indices; the first index gives the number within the sequence of the year in which a given flow occurs, whereas the second index gives a season number which runs cyclically from 1 to $m$. Thus the first index $i$ keeps account of an element's general position in the series. The index $j$ indicates which season in the set $\{1, \ldots, m\}$ is current. The quantities $\mu_j$ for $j$ running from 1 to $m$ are the mean values for flows in each of the $m$ seasons. Similarly, $\sigma_j$ is the standard deviation of flows in season $j$ and $\rho(j)$ is the correlation coefficient between flows in season $j$ and those in season $j - 1$. $\rho(j)$ should not be confused with $\rho_j$, the lag-$j$ serial correlation coefficient, or with $\rho^j$ (which is $\rho$ raised to the $j$th power and which can be proved to be equal to $\rho_j$ for the single season normal Markov process). $\rho(j)$ is a lag-one correlation coefficient, but it is restricted to adjacent pairs of flows from seasons $j - 1$ and $j$. Thus $\rho(j)$ is defined as

$$\rho(j) = \{E[x_{.,j} - \mu_j)(x_{.,j-1} - \mu_{j-1})]\}/(\sigma_j \, \sigma_{j-1}) \qquad (52)$$

Here $x_{.,j}$ means a flow from season $j$. $\rho(1)$ is the correlation coefficient between flows in the $m$th and first seasons, since the first season is assumed to follow the $m$th, so that when $j = 1$, $\mu_{j-1} = \mu_0$ is set equal to $\mu_m$ and similarly for $x_{.,0}$ and $\sigma_0$. The generation equation for a multiseason Markov model becomes

$$q_{i,j} = \mu_j + \frac{\rho(j)\sigma_j}{\sigma_{j-1}} (q_{i,j-1} - \mu_{j-1}) + t_{i,j}\sigma_j \sqrt{(1 - \rho(j)^2 )} \qquad (53)$$

The synthetic flow for the $i$th interval, which occurs in season $j$, is the sum of three terms. First is the mean of flows for season $j$,

second is the difference between the previous flow (season $j - 1$)
and its mean multiplied by the regression coefficient $\rho(j)\ \sigma_j/\sigma_{j-1}$,
and third is $t_{i,j}$, which is normally distributed with zero mean
and unit standard deviation so that when it is multiplied by the
standard error the flows have the desired variances.

Finally, we consider a multiseason (not a multilag) Markov model
for flows that are distributed (almost) as gamma. Let $\mu_j$, $\sigma_j$, $\rho(j)$
be defined as above and also let $\gamma_j$ equal the skewness coefficient
for flows from season $j$. Then define $\gamma_{t,j}$ to be the skewness of the
random variates in season $j$:

$$\gamma_{t,j} = \frac{[\gamma_j - \rho(j - 1)^3\ \gamma_{j-1}]}{[1 - \rho(j)^2]^{1.5}} \tag{54}$$

$t_{i,\gamma,j}$ is the transformed random sampling variate based on a normal
standardized deviate $t_{i,j}$; then

$$t_{i,\gamma,j} = \frac{2}{\gamma_{t,j}} \left( 1 + \frac{\gamma_{t,j}\ t_{i,j}}{6} - \frac{\gamma_{t,j}}{36} \right)^3 - \frac{2}{\gamma_{t,j}} \tag{55}$$

where $t_{i,j}$ is normal with zero mean and unit standard deviation.
Now $t_{i,\gamma,j}$ will be distributed almost as gamma with mean zero, standard deviation 1, and skewness $\gamma_{t,j}$. If $t_{i,\gamma,j}$ is used instead of
$t_{i,j}$ in the equation for the multiseason normal Markov model, the
result is a multiseason Markov model for flows, each season of
which is distributed almost as gamma:

$$q_{i,j} = \mu_j + \frac{\rho(j)\sigma_j}{\sigma_{j-1}} (q_{i,j-1} - \mu_{j-1}) + t_{i,\gamma,j}\ \sigma_j \sqrt{(1 - \rho(j))^2} \tag{56}$$

Again, because skewness varies from season to season, the represen-
tation is not statistically pure; sums of gamma variates are not
necessarily gamma unless certain restrictive conditions are met.

Empirical studies have used Markov generating models for stream-flow synthesis (Fiering, 1967). The work clearly indicates that the Markov model is more realistic than the independent normal model considered previously. It is, in fact, an inclusive model because it degenerates to the independent model if $\rho = 0$. (Actually the model is neither independent or normal, but uses and reproduces flows which are. This usage appears throughout this document.) For example, in the study of the range of cumulative departures described briefly in the section on statistics of the flow distributions, the Markov model predicts storage requirements that grow more rapidly with time than do the requirements predicated by an independent model. The empirical data of Hurst gave even more rapid growth with time, however, so that whereas the Markov model is an improvement over earlier models it does not yet give a completely satisfactory explanation of all features of flow patterns. The model does have several virtues. It is much easier to handle mathematically than are multilag models (considered below), particularly when flows are generated at several nearby sites in a basin. Moreover, the Markov model does reproduce some of the observed persistence in flows. The planner must decide whether the Markov model is adequate or whether a more complicated multilag model should be used. The design parameters may be evaluated adequately with a Markov scheme in many cases.

### Summary of Calculations for Markov Models

Each of these schemes generates a sequence of flows (or logarithms of flows) $q_1$, $q_2$, .... It is necessary to initiate the sequence with some value $q_1$ that will affect significantly the early flows in the sequence. Thus it is important not to use these early values in evaluating designs unless performance under transient conditions is at issue. We advise selecting any convenient value (perhaps $\bar{x}$) for $q_1$ and then discarding the first 50 flows in the sequence. The first accepted flow value will be far enough into the sequence that the effect on it of the starting condition will be negligible, and the extra computer time required to simulate a

random start is a matter of microseconds.  However, in complex systems, it is inadequate to guarantee that only the flows are begun at
random.  Reservoir levels and other state variables are to be ini-
tialized to a random starting position, and the warm-up period involves assignment of arbitrary values to the state variables and
then a substantial period of system operation with evaluation of economic or performance criteria.

*Single Season (Annual) Models*

 *Normally distributed flows.*  Use the mean $\bar{x}$, sample standard deviation $s$, and sample first serial correlation coefficient $r_1$ of the
historical flows as described in the section on the summary of calculations for estimating parameters and selecting a distribution.
Let $q_0 = \bar{x}$.  Let $t_1$, $t_2$, ... be a sequence of independent normally
distributed random numbers with mean 0 and standard deviation 1.
Then

$$q_1 = \bar{x} + r_1(q_0 - \bar{x}) + t_1 \, s\sqrt{(1 - r_1^2)}$$

$$= \bar{x}(1 - r_1) + r_1 \, q_0 + t_1 \, s\sqrt{(1 - r_1^2)} \tag{57}$$

$$q_2 = \bar{x}(1 - r_1) + r_1 \, q_1 + t_2 \, s\sqrt{(1 - r_1^2)}$$

and, in general,

$$q_i = \bar{x}(1 - r_1) + r_1 \, q_{i-1} + t_i \, s\sqrt{(1 - r_1^2)}$$

If $q_j$ is negative for some $j$, use the negative $q_j$ in the equation
for $q_{j+1}$ and discard $q_j$ without using it as a flow in the simulation.
Values of the mean, standard deviation, and first order serial correlation coefficient will differ between the historical and synthetic
flow sequences as a result of setting negative values of $q$ equal to
zero.  If the differences are unacceptably large, then a distribution
other than normal must be assumed.

 *Log normally distributed flows.*  To generate logarithms of flows,
first calculate $\bar{x}$, $s^2$, $g(x)$, and $r_1(x)$ from the historical record.

Solve equations for $a$, $\mu(y)$, $\sigma^2(y)$, and $\rho_1(y)$. Use these parameters for the $y$ in the normal generation scheme to produce a sequence $h_1$, $h_2$, ... of synthetic logarithms of flows. Then form a synthetic sequence of flows $q_1$, $q_2$, ... from

$$q_i = \exp(h_i) + a \tag{58}$$

Alternately, calculate the parameters $\bar{y}$, $s^2(y)$, and $r_1(y)$ of the logarithms of historical flows. Use the generation procedure for normal values to form a sequence $h_1$, $h_2$, ... of synthetic logarithms. Then form a sequence of synthetic logarithms $q_1$, $q_2$, ... from

$$q_i = \exp(h_i) \tag{59}$$

Another procedure for generating log normal flows is used by the U.S. Army Corps of Engineers (Beard, 1965).

*Gamma distributed flows.* Calculate the mean $\bar{x}$, the standard deviation $s$, and the first serial correlation coefficient $r_1$ of the historical flows. Also calculate the sample skewness coefficient of the flows $g$. Let $q_0 = \bar{x}$ and let $t_1$, $t_2$, ... be a sequence of independent normally distributed random variables with mean 0 and standard deviation 1. Define $g_\xi$ as

$$g_\xi = \frac{1 - r_1^3}{(1 - r_1^2)^{1.5}} \, g \tag{60}$$

Then define the sequence $\xi_1$, $\xi_2$, ... by

$$\xi_i = \frac{2}{g_\xi} \left(1 + \frac{g_\xi t_i}{6} - \frac{g_\xi^2}{36}\right)^3 - \frac{2}{g_\xi} \tag{61}$$

$\xi_1$, $\xi_2$, ... will be approximately distributed as gamma with mean 0, standard deviation 1, and skewness $g_\xi$. Then

$$q_i = \bar{x} + r_1(q_o - \bar{x}) + \xi_1 \, s \sqrt{(1 - r_1^2)}$$

$$= \bar{x}(1 - r_1) + r_1 \, q_o + \xi_1 \, s \sqrt{(1 - r_1^2)} \tag{62}$$

$$q_2 = \bar{x}(1 - r_1) + r_1 \, q_1 + \xi_2 \, s \sqrt{(1 - r_1^2)}$$

and, in general,

$$q_i = \bar{x}(1 - r_1) + r_1 \, q_{i-1} + \xi_i \, s \sqrt{(1 - r_1^2)}$$

As an alternate scheme, calculate $\bar{x}$, $s^2(x)$, and $g(x)$ for the historical flows. Select the positive integer $m$ for which $(8/m)^{0.5}$ is closest to $g(x)$. Then use the normal generation scheme with zero mean, unit standard deviation, and some set serial correlation value to form a series $y_1$, $y_2$, $\ldots$. Form a series $z_1$, $z_2$, $\ldots$ from

$$z_i = \sum_{j=(i-1)m+1}^{im} y_j^2 \tag{63}$$

and finally form a sequence $q_1$, $q_2$, $\ldots$ of synthetic flows from

$$q_i = \bar{x} + s(x)[(2m)^{-0.5} \, z_i - (m/2)^{0.5}] \tag{64}$$

*Multiple Season Models*

*Normally distributed flows.* These models assume that the year is divided into $m$ seasons (wet and dry seasons, for example, or perhaps 12 months). Assume that the historical data is $x_{1,1}$, $x_{1,2}$, $\ldots$, $x_{1,m}$, $x_{2,1}$, $\ldots$, $x_{p,m}$; i.e., the first index gives the year in which a flow occurred while the second index runs cyclically between 1 and $m$. First calculate $\bar{x}_j$, the mean flow for season $j$ (for each $j$ from 1 to $m$) as

$$\bar{x}_j = \frac{1}{p} \sum_{k=1}^{p} x_{k,j} \tag{65}$$

(that is, add up all the flows for season $j$ and divide by $p$). Next calculate $s_j$, the standard deviation of flows for season $j$ (for each $j$ from 1 to $m$) by

$$s_j^2 = \frac{1}{p} \sum_{k=1}^{p} x_{k,j}^2 - \frac{1}{p(p-1)} \left( \sum_{k=1}^{p} x_{k,j} \right)^2 \tag{66}$$

Calculate $r(j)$, the coefficient of correlation between flows in seasons $j$ and $j-1$, as

$$r(j) = \frac{\sum_{k=1}^{p} x_{k,j} \, x_{k,j-1} - p \, \bar{x}_j \, \bar{x}_{j-1}}{s_j \, s_{j-1} \, (p-1)} \tag{67}$$

When 0 occurs as the second subscript of an $\bar{x}$ or as the subscript of an $\bar{x}$ or $s$, it should be replaced by $m$ and, for an $x$, the first subscript should be decreased by one. (Note that in the calculation of $r(1)$ the term $x_{1,1} \, x_{0,m}$ appears in the numerator. If a flow for the last season of the previous year $(x_{0,m})$ is in fact available, it is used here. If a flow is not available, the sum contains only $p - 1$ terms and should be multiplied by $p/(p-1)$.) Let $t_{1,1}$, $t_{1,2}$, ..., $t_{1,m}$, $t_{2,1}$, ... be a sequence of independent, normally distributed random numbers with mean 0 and variance 1. Set $q_{0,m} = \bar{x}_m$. Then the following generation scheme gives the synthetic flows:

$$q_{1,1} = \bar{x}_1 + \frac{r(1)s_1}{s_m} (q_{0,m} - \bar{x}_m) + t_{1,1} \, s_1 \sqrt{[1 - r(1)^2]}$$

$$q_{1,2} = \bar{x}_2 + \frac{r(2)s_2}{s_1} (q_{1,1} - \bar{x}_1) + t_{1,2} \, s_2 \sqrt{[1 - r(2)^2]} \tag{68}$$

and, in general,

$$q_{i,j} = \bar{x}_j + \frac{r(j)s_j}{s_{j-1}} (q_{i,j-1} - \bar{x}_{j-1}) + t_{i,j}\, s_j \sqrt{1 - r(j)^2}$$

whenever the index $j$ reaches $m + 1$, reset it to 1 and increase the index $i$ by one. Negative flows are treated in the same manner as in the case of single season (annual) models discussed above.

*Gamma distributed flows.* Calculate the seasonal means, standard deviations, and correlation coefficients as described above for multiseason normal models. Finally, calculate $g_j$, the skewness coefficient of the flows for season $j$ (for each $j$ from 1 to $m$) as

$$g_j = \frac{\dfrac{1}{p}\displaystyle\sum_{k=1}^{p} x_{k,j}^3 - 3s_j^2\,\bar{x}_j + \bar{x}_j^3}{s_j^3} \tag{69}$$

For each $j$ from 1 to $m$ define $g_{t,j}$ by

$$g_{t,j} = \frac{g_j - r(j-1)^3\, g_{j-1}}{[1 - r(j)^2]^{1.5}} \tag{70}$$

Then let $t_{1,1}, t_{1,2}, \ldots, t_{1,m}, t_{2,1}, \ldots$ be a sequence of independent normally distributed random numbers with mean zero and standard deviation 1. Let

$$t_{i,\gamma,j} = \frac{2}{g_{t,j}} \left(1 + \frac{g_{t,j}t_{i,j}}{6} - \frac{g_{t,j}}{36}\right)^3 - \frac{2}{g_{t,j}} \tag{71}$$

Here the index $i$ remains fixed for $m$ flows and is then increased by one while $j$ runs from 1 to $m$ and then repeats. Finally, set $q_{0,m} = \bar{x}_m$ and

$$q_{1,1} = \bar{x}_1 + \frac{r(1)s_1}{s_m} (q_{0,m} - \bar{x}_m) + t_{1,\gamma,1}\, s_1 \sqrt{1 - r(1)^2}$$

$$q_{1,2} = \bar{x}_2 + \frac{r(2)s_2}{s_1}(q_{1,1} - \bar{x}_1) + t_{1,\gamma,2}\, s_2\sqrt{[1 - r(2)^2]}$$

$$\vdots \tag{72}$$

$$q_{i,j} = \bar{x}_j + \frac{r(j)s_j}{s_{j-1}}(q_{i,j-1} - \bar{x}_{j-1}) + t_{i,\gamma,j}\, s_j\sqrt{[1 - r(j)^2]}$$

### Examples of Markov Generation Schemes

A few calculations using the parameters derived from Table 4 and the random numbers given in Table 3 are shown below. The calculations are presented in tabular form to avoid any confusion.

*Annual Flows, Lag-One Markov Model, Normal Distribution*

We use a lag-one Markov procedure to generate the following series of flows:

| $i$ | $q_i$ | $q_i - \bar{x}$ | $r_1(q_i - \bar{x})$ | $\bar{x} + r_1(q_i - \bar{x})$ | $t_i$ | $t_i s\sqrt{1-r_1^2}$ | $q_{i+1}$ |
|---|---|---|---|---|---|---|---|
| 0 | 588.80 | 0.00 | 0.00 | 588.80 | -0.523 | -83.60 | 505.20 |
| 1 | 505.20 | -83.60 | -31.62 | 557.18 | 0.611 | 97.66 | 654.85 |
| 2 | 654.85 | 66.05 | 24.98 | 613.78 | -0.359 | -57.38 | 556.40 |
| 3 | 556.40 | -32.40 | -12.26 | 570.54 | -0.393 | -62.82 | 513.73 |
| 4 | 513.73 | -75.07 | -28.39 | 560.41 | 0.084 | 13.43 | 573.83 |
| 5 | 573.83 | -14.97 | -5.66 | 583.14 | -0.931 | -148.81 | 434.33 |
| 6 | 434.33 | -154.47 | -58.42 | 530.38 | -0.027 | -4.32 | 526.06 |
| 7 | 526.06 | -62.74 | -23.73 | 565.07 | 0.798 | 127.55 | 692.63 |
| 8 | 692.63 | 103.83 | 39.27 | 628.07 | 1.672 | 267.26 | 895.32 |
| 9 | 895.32 | 306.52 | 115.92 | 704.72 | -1.077 | -172.15 | 532.57 |

and so on until a long sequence is completed. Here $\bar{x}$ = 588.8 is the sample mean, $r_1$ = 0.37819 is the sample lag-one serial correlation coefficient, $s$ = 172.667 is the sample standard deviation, the $t_i$ are standard normal sampling deviates, and the $q(i)$ are the synthetic flows.

*Annual Flows, Lag-One Markov Model, Log Normal Distribution*

We use the sample moments of the historical flow values to determine the parameters for the logarithms. Let $\mu_y$, $\sigma_y^2$, $\rho_y$, and $\alpha$ be the mean, variance, lag-one serial correlation coefficient, and lower bound of the logarithms. We use $\bar{x} = 588.8$, $s^2 = 29,813.9$, $r_1 = 0.3782$, and $g = 0.992$ in the calculations. First,

$$0.992 = g = \frac{\exp{(3\sigma_y^2)} - 3\exp{(\sigma_y^2)} + 2}{[\exp{(\sigma_y^2)} - 1]^{3/2}}$$

or, after some work,

$$\exp{(\sigma_y^2)} = 1.1041$$

and

$$\sigma_y^2 = 0.099$$

Next, we use $s^2$ and $\sigma_y^2$ to find $\mu_y$:

$$s^2 = \exp{[2(\sigma_y^2 + \mu_y)]} - \exp{(\sigma_y^2 + 2\mu_y)}$$

$$29,813.9 = \exp{[2(0.099 + \mu_y)]} - \exp{(0.099 + 2\mu_y)}$$

Finally, $\mu_y = 6.233$.
Then $r_1$ and $\sigma_y^2$ determine $\rho_y$ from

$$r_1 = \frac{\exp{(\sigma_y^2 \rho_y)} - 1}{\exp{(\sigma_y^2)} - 1}$$

$$0.3782 = \frac{\exp{(0.099\ \rho_y)} - 1}{1.1041 - 1}$$

$$(0.1041)(0.3782) + 1 = \exp{(0.099\ \rho_y)}$$

$$1.03933 = \exp{(0.099\ \rho_y)}$$

$$0.039 = 0.099\ \rho_y$$

$$0.394 = \rho_y$$

And, finally

$$\bar{x} = a + \exp(\sigma_y^2/2 + \mu_y)$$

$$588.8 = a + \exp(0.0495 + 6.233)$$

or

$$a = 51$$

Now we use a lag-one Markov procedure to generate logarithms with mean $\mu_x = 6.233$, standard deviation $= (0.099)^{0.5} = 0.3146$, and lag-one serial correlation $\rho_y = 0.394$. If $h_1$, $h_2$, ... is the series of synthetic logarithms, then $a + \exp(h_1)$, $a + \exp(h_2)$, ... is the series of synthetic flows

| $i$ | $h_i$ | $h_i - \mu_y$ | $\rho_y(h_i - \mu_y)$ | $\mu_y + \rho_y(h_i - \mu_y)$ | $t_i$ |
|---|---|---|---|---|---|
| 0 | 6.2330 | 0.0000 | 0.0000 | 6.2330 | -1.536 |
| 1 | 5.7888 | -0.4442 | -0.1750 | 6.0580 | -0.454 |
| 2 | 5.9267 | -0.3063 | -0.1207 | 6.1123 | 0.071 |
| 3 | 6.1328 | -0.1002 | -0.0395 | 6.1935 | -2.129 |
| 4 | 5.5779 | -0.6551 | -0.2581 | 5.0749 | 1.525 |
| 5 | 6.4159 | 0.1829 | 0.0721 | 6.3051 | 0.261 |
| 6 | 6.3805 | 0.1475 | 0.0581 | 6.2911 | 2.319 |
| 7 | 6.9618 | 0.7288 | 0.2871 | 6.5201 | 0.972 |
| 8 | 6.8012 | 0.5682 | 0.2289 | 6.4569 | 0.767 |
| 9 | 6.6787 | 0.4457 | 0.1756 | 6.4086 | -2.849 |

| $i$ | $t_i \sigma_y \sqrt{1-\rho_y^2}$ | $h_{i+1}$ | $\exp(h_{i+1})$ | $a + \exp(h_{i+1}) = q_{i+1}$ |
|---|---|---|---|---|
| 0 | -0.4442 | 5.7888 | 326.622 | 377.622 |
| 1 | -0.1313 | 5.9267 | 374.913 | 425.913 |
| 2 | 0.0205 | 6.1326 | 460.746 | 511.746 |
| 3 | -0.6157 | 5.5779 | 264.503 | 315.503 |
| 4 | 0.4410 | 6.4159 | 611.484 | 662.484 |
| 5 | 0.0755 | 6.3805 | 590.245 | 641.245 |
| 6 | 0.6706 | 6.9618 | 1055.494 | 1106.494 |
| 7 | 0.2811 | 6.8012 | 898.950 | 949.950 |
| 8 | 0.2218 | 6.6787 | 795.278 | 846.278 |
| 9 | -0.8239 | 5.5847 | 266.319 | 317.319 |

and so on for a long sequence of flows. (Note that in determining $\mu_y$, $\sigma_y^2$, $\rho_y$, and $a$, the calculation sequence used above is strongly recommended. If one tries to solve the four nonlinear equations simultaneously instead, with some iterative method, the convergence will generally be very slow because the basic variable $\sigma_y^2$ is determined in an equation with a comparatively small constant side $(g = 0.992)$.)

## Multilag Models

Because Markov models do not always generate sufficiently realistic streamflow sequences, we now consider elaborations of that model. These changes arise primarily to accommodate runoff conditions in which the groundwater aquifer stores water from season to season and then contributes a fraction of it each season to form part of the total runoff. The behavior of the groundwater storage is represented by a multilag model; a model with a long memory. Instead of assuming that all influence of past flows on the current flow is contained in the magnitude of one previous flow, we assume that more than one past flow matters; i.e., we assume a deterministic part for the time series of the form

$$d_i = \beta_o + \beta_1 q_{i-1} + \cdots + \beta_m q_{i-m} \tag{73}$$

with $m > 1$. Clearly, this model is not a phenomenological explanation of flows; the real flows are most likely generated in nature by a complex process that depends on many hydrologic and meteorologic phenomena, not merely on past flows. Hence the justification for our multilag models must be operational. We justify the models by showing that they are capable of reproducing important statistical characteristics of historical flow patterns.

Figure 11 shows the correlogram (a plot of $k$ versus the lag-$k$ correlation coefficient) for a normal Markov process. Figure 12 shows the same correlogram with a typical empirical correlogram superimposed. The empirical correlogram shows large changes in magnitude and frequent changes of sign; not at all like the smooth

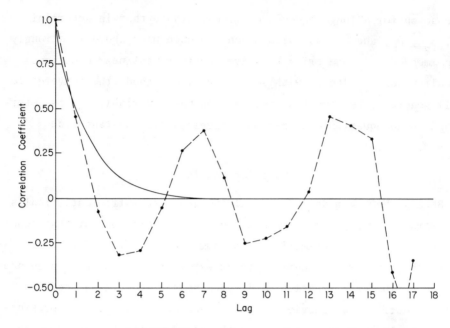

Fig. 12.  An empirical correlogram of annual flows (Eagle
   Creek, Indianapolis, Indiana) superimposed on theoretical
   correlogram.

Markovian correlogram that shows exponential decay.  Our justification for multilag models is that they are capable of producing the
jagged correlograms found in natural flow sequences.

The problem of how many previous flows to include in the model is
not an easy one to answer.  There are traditional statistical tests
of significance for correlation coefficients, but these tests are
really not appropriate here because they treat the historical flow
sequence $x_1$, ..., $x_n$ as circular in that $x_1$ is assumed to follow $x_n$.
This assumption is reasonable enough for a sequence of almost independent values, or for a sequence that is very much longer than the
longest lag being considered.  Persistence is important here, however, so that the magnitude of a flow is strongly affected by previous flows.  There is no reason whatever to assume that $x_1$ is a
likely follower of the sequence ... $x_{n-3}$, $x_{n-2}$, $x_{n-1}$, $x_n$.  For the
typically short streamflow records encountered in hydrology, such
end effects can be important.  Further, standard tests of significance assume what is called a null hypothesis; they then test the

hypothesis and assume that the error to be avoided most is rejection of the null hypothesis when it is in fact true. In tests of lagged correlation coefficients the null hypothesis is the assumption that a given coefficient is in fact zero, and the available tests are insensitive to economic consequences because they stringently protect against false rejection of the null hypothesis. Because one of the basic arguments for using operational hydrology is that economic consequences must be considered in the design process, standard tests are not appropriate for these results. Of course, this argument is a restatement of the traditional schism between 'hypothesis testers' and 'Bayesians', which is a familiar division in statistical decision theory.

We must then find a reasonable operational procedure for including lags; we adopt one that includes lags as long as it is practical and profitable to do so. (Richard Kronauer (personal communication, 1970) suggests a variety of theoretical results (based on large sample sizes) for converting moving average schemes into autoregressive form and for determining how many lags to include. Unfortunately, these results are not applicable here because hydrologic records are virtually always too short.) We start by determining estimates of $b_0$ and $b_1$ that give the best available approximation of $\beta_0$ and $\beta_1$ in the equation

$$x_i = \beta_0 + \beta_1 \, x_{i-1} \tag{74}$$

for the historical flows. That is, we fit a linear regression to historical flows and obtain the estimates $b_0$ and $b_1$. To measure the goodness of fit of the resultant linear model, we calculate the correlation coefficient (or really its square, the coefficient of determination) between the actual flow values $x_i$ and the values $b_0 + b_1 \, x_{i-1}$ which the model predicts for these flows. The coefficient of determination $R^2$ is

$$R^2 = \frac{\left[\left(\sum\limits_{i=2}^{n} x_i \hat{x}_i\right) - \left(\frac{1}{n-1}\right)\left(\sum\limits_{i=2}^{n} x_i\right)\left(\sum\limits_{i=2}^{n} \hat{x}_i\right)\right]^2}{\left[\sum\limits_{i=2}^{n} x_i^2 - \frac{1}{n-1}\left(\sum\limits_{i=2}^{n} x_i\right)^2\right]\left[\sum\limits_{i=2}^{n} \hat{x}_i^2 - \frac{1}{n-1}\left(\sum\limits_{i=2}^{n} \hat{x}_i\right)^2\right]} \qquad (75)$$

where $\hat{x}_i$ is defined as $b_0 + b_i x_{i-1}$. Most precoded regression programs on digital computers provide this $R^2$ value automatically. Next, we follow a similar procedure to obtain regression estimates $b_0$, $b_1$, and $b_2$ for the coefficients $\beta_0$, $\beta_1$, and $\beta_2$ in the model

$$x_i = \beta_0 + \beta_1 x_{i-1} + \beta_2 x_{i-2} \qquad (76)$$

and calculate a new $R^2$, the coefficient of determination between the calculated $\hat{x}_i$ values, $\hat{x}_i = b_0 + b_1 x_{i-1} + b_2 x_{i-2}$, and their observed counterparts $x_i$. $R^2$ measures the proportion of the variation in the flows that is explained by the regression, so changes in $R^2$ reflect changes in the explanatory power of the models. In most cases, the value of $R^2$ will increase or remain the same as we increase the number of lags. Therefore we require a more specific stopping rule. Typically the value of $R^2$ increases rapidly for the first few lags and remains on a relatively flat plateau (until it rises rapidly again for lags near the length of the historical record) (Figure 13). A practical stopping rule is to stop including lags when the $R^2$ values reach such a plateau. Thus our rule for including lags is to include an additional lag (include another antecedent flow in the model) if this lag produces an economically significant increase in the resultant value of $R^2$. Actually, we should include other stopping rules for the lags.

First, we note that because unavoidable inaccuracies such as round off errors accumulate during computer calculations, we may find $R^2 > 1$ for some models. The coefficient of determination cannot be greater than 1; certainly the model cannot explain more than the entire variation in flows. Impossible values for $R^2$ can arise

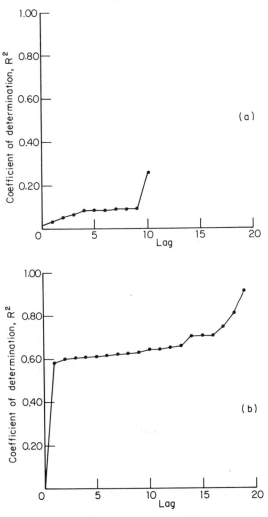

Fig. 13. Multiple correlograms.

in a perfectly well written program where they indicate not program errors but the accumulated inaccuracies inherent in machine calculation. If $R^2$ becomes larger than 1, it is mandatory to truncate the model by including no further lags. Second, even though we cannot give a satisfactory phenomenological explanation in terms of basin mechanics, it does seem reasonable to put some upper limit on the number of lags that may be included. The inclusion of flows from 35 years ago as explanatory variables for today's flows just does not seem reasonable; the aquifer's capacity to contribute runoff cannot survive that long.  An upper limit of 8 or 10 previous flows seems generous and certainly not restrictive.  Moreover, there is an excessive computational load involved with multisite models if the number of lags is at all large.  Third, we note that is dangerous to include lags that approach the length of the historical record because estimates of the correlations, based on limited date samples, are highly unstable and imprecise.

With respect to the random component, we assume that flows are normally distributed; alternately, we could be generating the logarithms of flows so that the flows themselves would be log normal. Let $t_1$, $t_2$, ... be a series of independent normal deviates with mean zero and unit standard deviation.  Then the multilag generation equation is

$$q_i = b_0 + b_1 q_{i-1} + b_2 q_{i-2} + \ldots + b_m q_{i-m} + t_i \, s\sqrt{1 - R^2} \qquad (77)$$

where $s$ is the standard deviation of the historical flow sequence. The factor $s\sqrt{1 - R^2}$ following the $t_i$ is the standard error of estimate, a scaling factor that preserves the desired variance; an analogous factor appears in the single lag Markov generation equation.

In some cases the modeler might want to use a multilag gamma model.  Unfortunately, no general theory for multivariate gamma variables is available and consequently the general use of multilag gamma models is not possible.  It is possible, however, to use the gamma distribution for one lag because there is a bivariate gamma distribution available.  Fiering (1967) discusses the theory required for such models; we do not reproduce those results here.

### Summary of Calculations for Multilag Models

1.  Because the gamma family is not suitable for multilag studies, we consider only normally distributed flows here.  The log normal distribution can also be used; in that case, the generation scheme produces logarithms of flows and exponentiation (anti-logs) gives the flows.

As in the Markov models, the flow generator must be primed with initial flow values; these values influence the first portion of the generated sequence. Consequently, it is wise to discard the first few generated flows. If 50 or 100 flow values are discarded, the first synthetic flow actually used in the analysis is far enough into the sequence that the effect of the arbitrarily chosen initial flows are negligible.

2. To use a multilag model, first specify rules for stopping the process of including lags in the model. $R^2$ is defined as the coefficient of determination (the square of the correlation coefficient) between historical flow values $x_i$ and the values $\hat{x}_i$ which the model predicts for those flows. Thus if the current model is of the form $x_i = \beta_0 + \beta_1 x_{i-1} + \dots + \beta_j x_{i-j}$ (i.e., the model contains $j$ lags) then first we use regression on the historical data to find the estimates $b_0, \dots, b_j$ of the coefficients. Then we form estimates of the $(j + 1)$st and all subsequent historical flows $(\hat{x}_{j+1}, \hat{x}_{j+2}, \dots, \hat{x}_n)$ as follows

$$\hat{x}_{j+k} = b_0 + b_1 x_{j+k-1} + \dots + b_j x_k \tag{78}$$

for $k = 1, 2, \dots, n - j$, from which

$$R^2 = \frac{\left[\sum\limits_{i=j+1}^{n} x_i \hat{x}_i - \frac{1}{n-j}\left(\sum\limits_{i=j+1}^{n} \hat{x}_i\right)\left(\sum\limits_{i=j+1}^{n} x_i\right)\right]^2}{\left[\sum\limits_{i=j+1}^{n} x_i^2 - \frac{1}{n-j}\left(\sum\limits_{i=j+1}^{n} x_i\right)^2\right]\left[\sum\limits_{i=j+1}^{n} \hat{x}_i^2 - \frac{1}{n-j}\left(\sum\limits_{i=j+1}^{n} \hat{x}_i\right)^2\right]} \tag{79}$$

$R^2$ is the proportion of the variation of the historical flow that is explained by the (linear regression) model. A reasonable rule is to stop including antecedent flows in the model if the last model obtained gave a value of $R^2$ which was not significantly better than that for the previous model; a better value of $R^2$ is a larger one.

As noted above, however, the value of $R^2$ will rarely decrease as we add more lags. As a rule, the value of $R^2$ increases rapidly as we add the first few lags and then levels off and remains on a plateau until we reach lags near the length of the historical record (Figure 13). A good intuitive rule is to stop including lags when the graph of $R^2$ against the number of lags included levels off. For a more mathematically precise discussion of the problem of when to stop including terms, the reader should consult the regression theory literature (e.g., Draper and Smith, 1966). Second, stop including lags whenever the accumulated inaccuracies in the computation have caused a value of $R^2$ that is not physically possible (i.e., a value of $R^2 > 1$). Third, assign an upper limit to the number of lags that may be included; 8 or 10 is a reasonable limit, and clearly even this is too large a bound if multiple sites (and all the matrix manipulations thereby implied) are used.

   3. Next, start to fit regression models of the form

$$x_i = \beta_0 + \beta_1 x_{i-1} + \ldots + \beta_m x_{i-m} \tag{80}$$

to the historical flows. Use regression analyses to obtain $b_0$ and $b_1$, estimates of $\beta_0$ and $\beta_1$ in the model

$$x_i = \beta_0 + \beta_1 x_{i-1} \tag{81}$$

Here the $x_i$ are the values $x_2$, ..., $x_n$ and the $x_{i-1}$ are $x_1$, ..., $x_{n-1}$ where $x_1$, ..., $x_n$ is the historical flow sequence. Calculate the $\hat{x}_i$, which are the values predicted by the model for $i = 2, \ldots, n$

$$\hat{x}_i = b_0 + b_1 x_{i-1} \tag{82}$$

Then calculate the coefficient of determination:

$$R^2 = \frac{\left[ \displaystyle\sum_{i=2}^{n} x_i \hat{x}_i - \frac{1}{n-1} \left( \displaystyle\sum_{i=2}^{n} x_i \right) \left( \displaystyle\sum_{i=2}^{n} \hat{x}_i \right) \right]^2}{\left[ \displaystyle\sum_{i=2}^{n} x_i^2 - \frac{1}{n-1} \left( \displaystyle\sum_{i=2}^{n} x_i \right)^2 \right] \left[ \displaystyle\sum_{i=2}^{n} \hat{x}_i^2 - \frac{1}{n-1} \left( \displaystyle\sum_{i=2}^{n} \hat{x}_i \right)^2 \right]} \tag{83}$$

Next, calculate by least squares $b_0$, $b_1$, and $b_2$, estimates of $\beta_0$, $\beta_1$, and $\beta_2$ in the model

$$x_i = \beta_0 + \beta_1 x_{i-1} + \beta_2 x_{i-2} \tag{84}$$

Determine the $\hat{x}_i$ values for this model, for $i = 3, 4, \ldots, n$

$$\hat{x}_i = b_0 + b_1 x_{i-1} + b_2 x_{i-2} \tag{85}$$

Calculate the new coefficient of determination and use the stopping criteria described above to decide whether or not to continue adding lags. When one of the stopping rules finally indicates an end to the model formulation, the result will be a set of regression coefficients $b_0$, $b_1$, $\ldots$, $b_m$ and a coefficient of determination $R^2$ for the corresponding model.

4. These coefficients and the $R^2$ value are then used to generate synthetic streamflows. Calculate the standard deviation of the historical flow sequence. Select $m$ initial flow values; $q_1 = q_2 = \cdots = q_m = \bar{x}$, the mean historical flow, would be a good choice. Let $t_{m+1}$, $t_{m+2}$, $\ldots$ be a sequence of normally distributed random numbers with mean 0 and standard deviation 1. Then generate the sequence

$$q_{m+1} = b_0 + b_1 q_m + b_2 q_{m-1} + \cdots + b_m q_1 + t_{m+1} \, s\sqrt{1 - R^2}$$

$$q_{m+2} = b_0 + b_1 q_{m+1} + b_2 q_m + \cdots + b_m q_2 + t_{m+2} \, s\sqrt{1 - R^2} \tag{86}$$

and, in general,

$$q_{m+i} = b_0 + b_1 q_{m+i-1} + b_2 q_{m+i-2} + \cdots + b_m q_i + t_{m+i} \, s\sqrt{1 - R^2}$$

If a negative flow is generated, use that flow in calculating the succeeding $m$ flows, but do not use that flow in the simulation.

Example of Calculations for a Multilag
Model, Normal Distribution

We now use the parameters derived from Table 4 and the random numbers in Table 3 to demonstrate the calculations for multilag models.

We begin with a lag-one model of the form

$$x_i = \beta_0 + \beta_1 \, x_{i-1}$$

and use linear regression to find the best (in the sense of least squares fit) estimates $b_0$ and $b_1$ from the historical flows.  The results are

$$b_0 = 378.28633$$

$$b_1 = 0.37007$$

Next, for $i = 2, \ldots, 20$ we form the estimates

$$\hat{x}_i = 378.28633 + 0.37007 \, x_{i-1}$$

and then calculate the coefficient of determination

$$R^2 = \frac{\left( \sum\limits_{i=2}^{20} x_i \, \hat{x}_i - \frac{1}{19} \sum\limits_{i=2}^{20} x_i \sum\limits_{i=2}^{20} \hat{x}_i \right)^2}{\left[ \sum\limits_{i=2}^{20} x_i^2 - \frac{1}{19}\left(\sum\limits_{i=2}^{20} x_i\right)^2 \right]\left[ \sum\limits_{i=2}^{20} \hat{x}_i^2 - \frac{1}{19}\left(\sum\limits_{i=2}^{20} \hat{x}_i\right)^2 \right]}$$

$$= \frac{\left[ 6853726 - 1/19 \, (11347)(11347.005) \right]^2}{\left[ 7316131 - 1/19 \, (11347)^2 \right]\left[ 6853729 - 1/19(11347.005)^2 \right]}$$

$$= \frac{5956033744}{(539583)(77175)} = 0.14303$$

The value of $R$ is then 0.3782, the value found for $r_1$ for the sample data. This result is expected since the lag-one model

$$x_i = \beta_0 + \beta_1 x_{i-1}$$

is a restatement of the lag-one Markov model.

The quantities $b_0$, $b_1$, and $R$ are now used to generate a synthetic flow sequence

| | $q_{i-1}$ | $b_1 q_{i-1}$ | $b_0 + b_1 q_{i-1}$ | $t_i$ | $t_i s \sqrt{1 - R^2}$ | $q_i$ |
|---|---|---|---|---|---|---|
| 1 | 588.800 | 217.896 | 596.182 | -0.121 | -19.3410 | 576.841 |
| 2 | 576.841 | 213 471 | 591.757 | 0.968 | 154.7276 | 746.484 |
| 3 | 746.484 | 276.250 | 654.536 | -1.943 | -310.5741 | 343.962 |
| 4 | 343.962 | 127.289 | 505.576 | 0.581 | 92.8685 | 598.444 |
| 5 | 598.444 | 221.465 | 599.751 | -0.711 | -113.6481 | 486.103 |
| 6 | 486.103 | 179.891 | 558.178 | -0.060 | -9.5906 | 548.587 |
| 7 | 548.587 | 203.015 | 581.301 | -0.482 | -77.0441 | 504.257 |
| 8 | 504.257 | 186.609 | 564.896 | -0.746 | -119.2426 | 445.653 |
| 9 | 445.653 | 164.922 | 543.208 | -0.747 | -119.4024 | 423.806 |
| 10 | 423.806 | 156.837 | 535.123 | 1.254 | 200.4426 | 735.566 |

and so on.

We next consider a lag-two model

$$x_i = \beta_0 + \beta_1 x_{i-1} + \beta_2 x_{i-2}$$

The linear regression estimates of $\beta_0$, $\beta_1$, and $\beta_2$ are

$$b_0 = 373.3599$$
$$b_1 = 0.2673$$
$$b_2 = 0.1286$$

Then for $i = 3, \ldots, 20$

$$\hat{x}_i = 373.3599 + 0.2673 x_{i-1} + 0.1286 x_{i-2}$$

The coefficient of determination is

$$R^2 = \frac{\left( \sum\limits_{i=3}^{20} x_i \hat{x}_i - \frac{1}{18} \sum\limits_{i=3}^{20} x_i \sum\limits_{i=3}^{20} \hat{x}_i \right)^2}{\left[ \sum\limits_{i=3}^{20} x_i^2 - \frac{1}{18} \left( \sum\limits_{i=3}^{20} x_i \right)^2 \right] \left[ \sum\limits_{i=3}^{20} \hat{x}_i^2 - \frac{1}{18} \left( \sum\limits_{i=3}^{20} \hat{x}_i \right)^2 \right]}$$

$$= \frac{[6748984.5 - 1/18 \ (10971)(10971.00002)]^2}{[7174755 - 1/18 \ (10971)^2][6748948.5 - 1/18 \ (10971.00002)^2]}$$

$$= \frac{3863860417}{(48793)(62160)} = 0.1274$$

This value of $R^2$ is smaller than the one obtained for the lag-one model so we do not include more than one lag. (The reader familiar with linear regression theory will probably object to our assertion that we add another lag and find that $R^2$ actually decreases, because adding more variables to a regression model cannot decrease the explanatory power of the model. The resolution of this apparent contradiction is that we do not simply add another variable but also modify the model slightly. In the lag-one regression we use 19 pairs of flows ($x_{i-1}$, $x_i$ for $i = 2, \ldots, 20$); thus, we estimate the 19 flow values $x_2, \ldots, x_{20}$ to find $R^2 = 0.143$. In the lag-two model we can estimate only 18 values $x_3, \ldots, x_{20}$ to find $R^2 = 0.127$. The second model is not an extension of the first, but is instead an extension of the lag-one model in which only 18 flows $x_3, \ldots, x_{20}$ are predicted from their predecessors $x_2, \ldots, x_{19}$; that 18-point lag-one model gives $R^2 = 0.111$ and, as predicted by theory, the $R^2$ value of 0.127 for the lag-two model is at least as good as 0.111.) If the lag-two model $R^2$ were larger, we would generate synthetic flows from

$$q_i = 373.3599 + 0.2673 q_{i-1} + 0.1286 q_{i-2} + t_i \ (172.667) \sqrt{1 - 0.1274}$$

where we would initialize the first two flows to 588.8.

# 5    OTHER CONSIDERATIONS IN THE USE OF OPERATIONAL HYDROLOGY

## Multisite Models

All the models considered so far deal with flows for a single site, whereas many actual planning projects involve a system of sites such as a system of reservoirs and appurtenant structures. Several new problems arise in the streamflow generation when several sites are used. It is not satisfactory simply to use single site generation procedures for each of the sites in turn, because flows at the various sites can be strongly interrelated. If a particular month is unusually wet at one site in an area, it is very likely that the same month will be wet at nearby sites. Moreover, if flow is high at one time on a particular stream, then it will tend to be high sometime later in a lower reach of the stream. Independent generation of the flow values for multiple sites cannot preserve spatial and temporal correlations between flows; consequently, multivariate techniques are needed (Matalas, 1967; Fiering, 1964).

Unfortunately, the mathematics involved in the multisite model is far more complicated than that involved in the other models considered here. A brief outline of the calculations for multisite models appears in the appendix. The planner should consider using an existing computer program (Pisano, 1968) for multisite basin simulation; that program is available through the Federal Water Quality Administration.

As discussed in the appendix, one comparatively simple form of the multisite model assumes that all the connection among flows at related sites can be adequately summarized in relations among the random components of flows at the sites. Thus, when using that model, one synthesizes correlated random components and then uses them

in Markovian generating equations. The use of such correlated random components is reasonable in that the random component summarizes
all that we do not know about the flow at a given site; the unknown
factors affecting flow at one site are assumed correlated with the
unknown factors at related sites.

## Refinements of the Generation Schemes

There are several modifications that can easily be made in these
generation schemes so as to produce flow sequences for special purposes. In some regions it is not entirely appropriate to use uni-
modal (single-peaked) probability distributions such as the normal,
log normal, or gamma. In some arid regions the precipitation distributions have two peaks; i.e., probably there will be either no
rain or an intense cloudburst, but intermediate amounts of rain are
infrequent. In the northeastern United States there seem to be two
populations for precipitations (or for streamflows). First, there
is the usual population (perhaps log normal) of flows, and second,
there is the population of extreme flows caused by hurricanes and
extra tropical coastal storms. If flows caused by hurricanes are
combined with the other flows to determine overall statistics of
flows, the result will be a population intermediate between the
usual and extreme populations, not at all representative of actual
flow patterns. Use of such an intermediate or smoothing population
might distort simulation results quite seriously, so it is better to
deal with two flow populations. One determines the statistical parameters of the usual flows and selects a model for generating them
and similarly determines the parameters of the flows caused by hurricanes and selects a model for generating them. Then, the probability of a hurricane is found; perhaps one would use a monthly model and fix the probability of a hurricane for each month of the
year. If $\Phi$ were the probability of a hurricane during month $i$, then
the hurricane population would be used for flows for that month with
probability $\Phi$ and the usual population would be used with probability $1 - \Phi$. This scheme is shown in Figure 14.

Another refinement of the generation model allows synthesis of
reasonably consistent flows for time intervals that are subdivisions

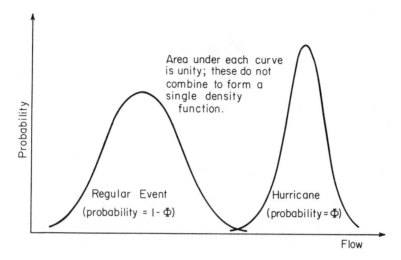

Fig. 14. Combined storm distributions.

of one another (Harms and Campbell, 1967). Assume that the planner is interested in both annual and monthly flows for the same site. He would examine the historical patterns of both monthly and annual flows, determine their parameters, and select distributions for them. The planner might find, for example, that annual flows are distributed approximately normally whereas monthly flows seem to derive from the log normal family. Assume that he decides to use a Markov generation scheme. The problem is then to generate normal annual flows and log normal monthly flows that preserve both the historical correlation between annual flows and the correlations between flows in successive months. Harms and Campbell suggest the following partial solution to the problem. First, generate a sequence of annual flows $q_1$, $q_2$, ... from the desired normal population that show the desired annual serial correlation value. Second, generate a tentative set $q_{1,i}^o$, $q_{2,i}^o$, ..., $q_{12.i}^o$ of monthly flows from the desired log normal population for year $i$ showing the month by month correlations of the historical record. To do this simply use the multiseason log normal Markovian model described in previous sections. Third, scale the monthly values to make the monthly flows sum to the annual flow $q_i$. If each monthly flow gives the total flow for the month, then the total of the monthly flows is

$$\sum_{j=1}^{12} q^{o}_{j,i} = Q \qquad (87)$$

Let the final monthly flows for year $i$ be defined by

$$q_{j,i} = \left(\frac{q^{o}_{j,i}}{Q}\right) q_i \qquad (88)$$

Thus $q^{o}_{j,i}/Q$ gives the proportion of the yearly flow that occurs in month $j$; $q_i$ is the annual flow for year $i$.

This weighting scheme produces normal annual flows, log normal monthly flows, monthly flows that sum to the annual flow, and monthly ly flows whose serial correlations within a year have the historical values as expected values. Because the flows in adjacent years are subjected to different weighting values, the correlation coefficient between flows in the last month of one year and flows in the first month of the following year are not preserved. To minimize the effects of this problem, it is wise to select the end of the water year at a relatively noncrucial time; not, for example, in the middle of the dry season for a study that focuses on low flows.

Fiering (unpublished memorandum, 1970) offers another approach to the problem of bringing annual and seasonal correlations into harmony. This approach is described in the section on refinements in evaluating sample statistics.

### Game and Decision Theories

A description of the tests for determining the distribution of flows is given in the section on selecting a distribution. We note in that section that the historical record alone often does not indicate an unambiguous choice of distribution. Frequently the planner finds that two distributions seem reasonable. In such cases, decision and game theories offer further aid in selecting a distribution. These disciplines suggest that, if no clear decision is

dictated by the previous analysis, the planner should invoke economic considerations in the further selection. Thus in a simple example the planner may have narrowed the alternatives to include only
the normal and log normal distributions. Next he makes two design
plans, both using streamflow synthesis and system simulations. One
design plan ($\Delta_1$, say) is best under the assumption of normal flows
and the second ($\Delta_2$) is best if flows are log normal. The planner
must have an objective function for identifying the optimal designs
under each of the two assumptions. The next step is to test the consequences, in terms of the objective function, of accepting design $\Delta_1$
if the true flow distribution is not normal but log normal and also
the results of accepting design $\Delta_2$ when the actual flow distribution
is normal instead of log normal. These results are obtained by routing appropriate sets of synthetic flows through both 'optimal' system
designs, just as the designs are evaluated in the first place through
simulation. Let $\Theta_1$ signify that flows are normally distributed and
$\Theta_2$ signify that flows are log normal. Similarly, let $\Phi_1$ be the assumption that the flows are normal (with resulting optimal design $\Delta_1$)
and let $\Phi_2$ be the assumption that flows are log normal (with resulting design $\Delta_2$). Then $\beta_{ij}$ is the expected benefit if assumption $\Phi_i$ is
adopted when $\Theta_j$ is the true condition, and the information obtained
by all four simulations can be summarized in the following payoff matrix or 2 x 2 contingency table:

| Design Assumptions | Flow $\Theta_1$ | Flow $\Theta_2$ | Optimal Design |
|---|---|---|---|
| $\Phi_1$ | $\beta_{11}$ | $\beta_{12}$ | build plan $\Delta_1$ |
| $\Phi_2$ | $\beta_{21}$ | $\beta_{22}$ | build plan $\Delta_2$ |

Plan $\Delta_1$ is optimal for normal flows and plan $\Delta_2$ is optimal for
log normal flows.

The planner may wish to use a slightly more complicated payoff matrix;
he may wish to consider another possible state of nature $\Theta_3$, which
corresponds to any distribution other than normal or log normal. It

is not easy to generate synthetic flow sequences for $\theta_3$, because
'other' distribution is not specified.  The analyst might use permu-
tations of the historical flow sequence or some critical period of
the record as an approximation to sequences generated when $\theta_3$ holds.
These sequences for $\theta_3$ are then routed through systems using designs
$\Delta_1$ and $\Delta_2$ to obtain the performance indices $\beta_{13}$ and $\beta_{23}$.  The payoff
matrix is now

| Design Assumptions | Flow $\theta_1$ | Flow $\theta_2$ | Flow $\theta_3$ | Optimal Design |
|---|---|---|---|---|
| $\Phi_1$ | $\beta_{11}$ | $\beta_{12}$ | $\beta_{13}$ | build $\Delta_1$ |
| $\Phi_2$ | $\beta_{21}$ | $\beta_{22}$ | $\beta_{23}$ | build $\Delta_2$ |

In a few fortunate cases the choice is extremely simple.  $\beta_{11}$
must be at least as large as $\beta_{21}$ since $\Delta_1$ is chosen as the best design if $\theta_1$ holds.  If, in addition, $\beta_{12}$ is as large as $\beta_{22}$ and $\beta_{13}$
is as large as $\beta_{23}$, then the decision problem is said to have a dom-
inant solution*.  $\Delta_1$ is optimal regardless of the true distribution
of flows and consequently is the best design.  Similarly, the decision problem (or game*) has a saddle point if $\Delta_2$ is optimal, regardless of the actual state of nature (or flow distribution).  In most
cases, however, a simple pure strategy* is not available and the
planner has to apply the theory of games to obtain a solution.
(Terms marked thus (*) are standard terms in game theory.  The interested reader can refer to any number of useful books on the subject,
but a particularly lucid exposition is found in *The Compleat
Strategyst* (Williams, 1954).)  The game theory solution to the decision problem is a pair of probabilities, $p_1$ and $p_2$, which have the
property that the choice of $\Delta_1$ with probability $p_1$ and the choice of
$\Delta_2$ with probability $p_2 = 1 - p_1$ provide a return (benefits) equal to
the value of the game, no matter what the probabilities of occurrence
of the various flow distributions*.  (This conclusion obtains under
the pessimistic objective function known as the Wald criterion.
Other criteria lead to different solutions (e.g., the criteria of

Laplace, Hurwicz, and Savage.) If the decision maker repeats this decision many times (i.e., the same problem keeps recurring) then the optimal strategy is to choose $\Delta_1$ and $\Delta_2$ with probabilities $p_1$ and $p_2$, respectively. Generally, however, the planner faces a one-time decision problem, whereupon he should consider the use of a weighted sum of the two decisions

$$\Delta^* = p_1 \, \Delta_1 + p_2 \, \Delta_2 \tag{89}$$

Use of such a weighted average of the two designs is justified on the implicit assumption that the benefit function is linear (or, at least, reasonably close to linear) between the two (or, indeed, any number of) design choices. One certainly would not want to make an intermediate choice worse than both $\Delta_1$ and $\Delta_2$ regardless of the state of nature. If $\Delta_1$ indicates a very large value of some design parameter while $\Delta_2$ indicates a very small value, it might well be that an intermediate value would always be bad and, hence, should not be used. Further, the use of a weighted average assumes that intermediate designs such as $\Delta^*$ are possible; if only a few designs are feasible then the choice obviously must be limited to feasible designs.

There are various other decision-theoretic criteria for choosing among alternatives. Charnoff and Moses (1959) provide particularly readable discussions of various types of decision rules; Luce and Raiffa (1957) give a more thorough discussion of game theory and decision making.

## Case Histories

The first application of stochastic streamflow simulation to a real situation was undertaken in connection with a study by the Meramec River Basin Commission (1962). It was a modest simulation program, at least by present standards, and was integrated into the planning of a few small reservoirs in the St. Louis area.

The next effort was made in response to a request from Pakistan's president, Ayub Khan. President Kennedy directed his Scientific Advisory Council to undertake a technical assistance program to

relieve waterlogging and salinity of agricultural lands in the
Punjab region of West Pakistan.  To evaluate the technological response of the proposed remedies, a group at Harvard simulated a well
field of heroic proportions and generated synthetic traces of river
flow, irrigation diversions, and rainfall.  The output comprises a
trace of groundwater depth versus time and an evaluation of the economic benefits of increased agricultural productivity.  A near optimal design was recommended and implementation has begun (Fiering,
1965).

Members of that same group studied the disposal of radioactive
wastes in streams by simulating hydrologic events and the introduc-
tion of radioactive pollutants.  The technologic function is such
that during periods of low flow the pollutant is deposited in the
stream bank while during periods of high flow the radioactive material is removed by scouring action and is carried downstream toward
the zone of water use.  Meanwhile, both in transit and in bank storage, the material is subject to exponential decay governed by the
half-life of the pollutant.  Any attempt to deduce the concentration
of radioactive flux at downstream use points by analytical techniques would be intractable, but stable estimates of the parameters
of the concentration distribution can be obtained by simulation with
long synthetic traces.  The analysis was applied to the Clinch River
in Tennessee (Thomas and Fiering, 1964).

A subsequent study dealt with the Delaware River basin and one of
its subbasins, the Lehigh River.  Economic data developed by the
U.S. Army Corps of Engineers (1962) in its summary report is used in
a simulation analysis of the Lehigh River that encompasses six reservoirs, nine turbines, water supply diversion works, and storage facilities for flood control and recreation.  The results indicate an
improvement of approximately 50% on the best of the Corps' designs,
and the study has been generalized to include the 42 reservoirs and
turbines of the entire Delaware basin (Hufschmidt and Fiering, 1966).

More recent studies include synthesis and simulation packages applied to the Potomac estuary (Davis, 1968), the Mississippi River

delta (Meta Systems Inc., 1969b), the Monongahela River (Meta Systems Inc., 1969a), the Ganges-Brahmaputra basin (work currently in progress at the Harvard Center for Population Studies), and a variety of other systems whose designs depend on various synthetic inputs such as hydrologic flows, precipitation, economic demands, traffic loads, and waste loads.

Thus what is offered here is a generalized procedure for inputs to decision models in a wide sense. Performance criteria can now be calculated in a statistically meaningful way, and thus a wide swath is cut through the confusion and ambiguity associated with traditional measures or indices of performance.

A typical example of such a study is the application of synthesis of flood stages and flood damages to the design of a national flood insurance program (Schaake and Fiering, 1967). There are probably many other examples in government and private practice that emanate from engineers throughout the world. It would be impossible to list all the examples here, and only a few of the earliest are given to suggest the range of application.

We note that the several cases cited emphasize a range of application, running from simple storage yield analysis to complicated water quality models that incorporate many sites, multiple objectives, seasonal and daily (or even shorter) time intervals, and so forth. Each application required some adjustment of the simple recursive equation or of the underlying assumption of a Markov process, but in these incremental steps the theory and art of flow synthesis moved forward.

# APPENDIX: MATHEMATICS FOR THE MULTISITE MODEL

Assume that there are $m$ sites for which synthetic streamflows are desired. We use a superscript to denote the site number of a particular flow or flow parameter. Thus $x_1^{(4)}$, $x_2^{(4)}$, ..., $x_n^{(4)}$ would be the flow values for the fourth station whereas $\bar{x}^{(4)}$, $s^{(4)}$, $g^{(4)}$, and $r_1^{(4)}$ would be the sample mean, standard deviation, skewness coefficient, and first serial correlation coefficient, respectively, for that station. The lag-zero cross correlation between the historical flows at stations $k$ and $l$ is

$$\frac{\frac{1}{n} \sum_{i=1}^{n} (x_i^{(k)} - \bar{x}^{(k)})(x_k^{(l)} - \bar{x}^{(l)})}{s^{(k)} \, s^{(l)}} \tag{A1}$$

and will be denoted $r_0^{(k,l)}$. We define the $(m \times 1)$ vector $x_i$ whose $l$th component is $x_i^{(l)} - \bar{x}^{(l)}$; $\bar{x}^{(l)}$ is the mean flow for station $l$. One means of generating multivariate synthetic flows is

$$x_{i+1} = A x_i + B_1 \, \varepsilon_{i+1} \tag{A2}$$

Here $\varepsilon_{i+1}$ is an $(m \times 1)$ matrix of random components that are assumed to be independent of $x_i$. $A$ and $B$ are both $(m \times m)$ matrices. Since values of flow for all stations during interval $i$ appear in the definition of flows for period $(i + 1)$, the flows are interrelated. The matrices $A$ and $B$ specify the precise form of the interdependence. We assume that $E(\varepsilon_{i+1})$, the expected value or mean of the vector of random components, is a zero vector, that the elements of $\varepsilon_{i+1}$ are independent of one another, and that the variance of each element of $\varepsilon_{i+1}$ is one. Thus

$$E(\varepsilon_{i+1}\ \varepsilon_{i+1}{}^T) = I \qquad (A3)$$

the $m \times m$ identity matrix. We also need the $m \times m$ matrix $M_0 = E(x_l\ x_l{}^T)$. The $l$th diagonal element of $M_0$ is $(S^{(l)})^2$ and the $(k,\ l)$ element for $k \neq l$ is

$$r_0{}^{(k,l)}\ S^{(k)}\ S^{(l)} \qquad (A4)$$

Thus

$$M_0 = \begin{bmatrix} (S^{(l)})^2 & r_0{}^{(1,2)}S^{(1)}S^{(2)} & \cdots & r_0{}^{(1,m)}S^{(1)}S^{(m)} \\ r_0{}^{(2,1)}S^{(2)}S^{(1)} & (S^{(2)})^2 & & \vdots \\ \vdots & & \ddots & \vdots \\ r_0{}^{(m,1)}S^{(m)}S^{(1)} & \cdots & & (S^{(m)})^2 \end{bmatrix}$$

$$(A5)$$

The $(m \times m)$ matrix $M_1$ is defined as $E(x_{i+1}x_i{}^T)$. The $l$th diagonal element of $M_1$ is $r_1{}^{(l)}\ (S^{(l)})^2$; for $k \neq l$ the $(k,l)$th element of $M_1$ is $r_1{}^{(k,l)}S^{(k)}S^{(l)}$ where $r_1{}^{(k,l)}$ is the lag-one correlation coefficient between flows at site $l$ and flows at site $k$. In many cases, the planner wishes to preserve the historical mean, standard deviation, and lag-one serial correlation coefficient for each site and also the lag-zero cross correlations between sites, but he does not care about the lag-one cross correlations between sites (the correlations between the flow at one site at a given time and the flow at another site for the next time period). If that is the case, the calculations for the multisite model can be simplified. In general

$$A = M_1\ M_0{}^{-1} \qquad (A6)$$

defines the matrix $A$ for the generation equation. If the lag-one cross correlations are not of interest, then the matrix $A$ may be defined as

$$
A = \begin{bmatrix}
r_1^{(1)} & 0 & 0 & 0 \cdots 0 \\
0 & r_1^{(2)} & 0 & 0 \cdots 0 \\
0 & 0 & r_1^{(3)} & 0 \cdots 0 \\
\cdot & \cdot & \cdot & \cdot \\
\cdot & \cdot & \cdot & \quad \cdot \\
\cdot & \cdot & \cdot & \qquad \cdot \\
0 & 0 & 0 & r_1^{(m)}
\end{bmatrix}
\tag{A7}
$$

In this case, the matrix $M_1$ is defined with diagonal elements equal to $r_1^{(l)} (S^{(l)})^2$ as before but with new off diagonal elements; the $(k,l)$th element is now $r_0^{(k,l)} r_1^{(l)}$. These new entries in the $M_1$ matrix are the true lag-one cross correlations for lag-one Markov processes. Thus the suggested simplification assumes a lag-one Markov model and insists on reproduction of the historical means, standard deviation, same site lag-one correlations, and lag-zero cross correlations; it then assigns the values predicted by the model to the lag-one cross correlations even though we know that the model will not hold exactly.

Next we determine the elements of the matrix $B$ by noting it is a solution of

$$
BB^T = M_0 - M_1 M_0^{-1} M_1^T
\tag{A8}
$$

where $M_1^T$ is the transpose of $M_1$. To solve for $B$ we determine the eigenvalues of the matrix $M_0 - M_1 M_0^{-1} M_1^T = M$. The techniques of principal component analysis (Kendall, 1961) are applied to the equation $BB^T = M$ to solve for $B$. The matrix $B$ will then satisfy

$$
B^T B = \lambda
\tag{A9}
$$

where $\lambda$ is a diagonal matrix whose elements are the eigenvalues of the matrix $M$. The matrix $B$ is not uniquely identified in this procedure.   Indeed, if θ is any matrix which satisfies

$$\theta\theta^T = I \tag{A10}$$

then $(B\theta)(B\theta)^T$ will equal $M$ and the matrix $B\theta$ can be used in the
generation process.   Since there will be many matrices that satisfy
$\theta\theta^T = I$, there will be many acceptable choices for the generation
scheme.

It is instructive to consider the form of the generation scheme
when we assume that the matrix $A$ is diagonal and we define the off
diagonal elements of $M_1$ as discussed above.   The difference between
the flow at site $l$ in time interval $t{+}1$ and the mean flow for site $l$
is defined by

$$x_{t+1}^{(l)} = r_1^{(l)} x_t^{(l)} + \sum_{s=1}^{m} b_{l,s} \, \varepsilon_{t+1}^{(s)} \tag{A11}$$

Thus the value of $x_{t+1}^{(l)}$ depends on the value of $x_t^{(l)}$ and also on
the random components vector for the time period $t{+}1$.   It depends on
the previous flow at the same site in a proportion given by the lag-
one correlation coefficient for that site; this part of the formation
is identical to that for the single site Markov model.   What is new
in this multisite formulation is that the random components for the
$m$ different sites at time $t{+}1$ are combinations of the elements of
the vector $\varepsilon_{t+1}$, thus reflecting the relations among the various
random components.   If the $B$ matrix is diagonal, the flows at the
different stations are not related and

$$x_{t+1}^{(l)} = r_1^{(l)} x_t^{(l)} + \varepsilon_{t+1}^{(l)} \tag{A12}$$

so that we simply generate Markovian flow values for $m$ completely
separate models, one for each site.   In the general case $B$ is not
diagonal and the random component of the flow at one station is related to the random component of the flow at another station.

The mathematics involved in the multisite models is considerably more complicated than the mathematics in any of the previous models. The reader is referred to the literature for further explanations of the problems and concepts involved. For a discussion of eigenvalues, see any text on linear algebra (Birkhoff and MacLane, 1953; Herstein, 1964; Noble, 1969). Ralston (1965) discusses the problem of determining eigenvalues on a computer; Kendall (1961) discusses principal component analysis; and Fiering (1964) and Matalas (1967) discuss multisite flow generation.

# REFERENCES

Aitchison, J., and J. A. C. Brown, *The Log-Normal Distribution*, 176 pp., Cambridge University Press, London, 1957.

Barnes, F. B., Storage required for a city water supply, *J. Inst. Eng.*, *26*, Australia, 198, 1954.

Beard, Leo R., Use of interrelated records to simulate streamflow, *J. Hydraul. Div.*, *Amer. Soc. Civil Eng.*, *91*, 13-22, 1965.

Benjamin, Jack R., and C. Allin Cornell, *Probability, Statistics, and Decision for Civil Engineers*, 684 pp., McGraw-Hill, New York, 1970.

Benson, M. A., and N. C. Matalas, Synthetic hydrology based on regional statistical parameters, *Water Resour. Res.*, *3*(4), 931-935, 1967.

Birkhoff, Garrett, and Saunders MacLane, *A Survey of Modern Algebra*, 472 pp., Macmillan, New York, 1953.

Box, G. E. P., and M. E. Muller, A note on the generation of normal deviates, *Ann. Math. Statist.*, *28*, 610-611, 1958.

Bryant, George T., Stochastic theory of queues applied to design of impounding reservoirs, Ph.D. dissertation, Harvard University, Cambridge, Massachusetts, 1961.

Chernoff, Herman, and Lincoln E. Moses, *Elementary Decision Theory*, 364 pp., John Wiley, New York, 1959.

Chow, Ven Te, Discussion of Hurst's 'Long-term storage capacity of reservoirs,' *Trans. Amer. Soc. Civil Eng.*, *116*(776), 800-802, 1951.

Close, Edward R., Leo R. Beard, and David R. Dawdy, Objective determination of safety factors in reservoir design, *J. Hydraul. Div.*, *Amer. Soc. Civil Eng.*, *96*, 1167-1177, 1970.

Crawford, Norman, and Ray K. Linsley, The synthesis of continuous streamflow hydrographs on a digital computer, 121 pp., *Dep. Civil Eng.*, *Tech. Rep. 12*, Stanford University, Stanford, California, July 1962.

Davis, Robert K., *The Range of Choice in Water Management, Resources for the Future*, 196 pp., The Johns Hopkins Press, Baltimore, Maryland, 1968.

Draper, N. R., and H. Smith, *Applied Regression Analysis*, 407 pp., John Wiley, New York, 1966.

Feller, William, The asymptotic distribution of the range of sums of independent random variables, *Ann. Math. Stat.*, *22*, 427-432, 1951.

Fiering, M. B, Multivariate techniques for synthetic hydrology, *J. Hydraul. Div.*, *Amer. Soc. Civil Eng.*, *90*, 43-60, 1964.

Fiering, M. B, Revitalizing a fertile plain, *Water Resour. Res.*, *1*(1), 41-61, 1965.

Fiering, M. B, *Streamflow Synthesis*, 139 pp., Harvard University Press, Cambridge, Massachusetts, 1967.

Harms, Archie A., and Thomas H. Campbell, An extension to the Thomas-Fiering model for the sequential generation of streamflow, *Water Resour. Res.*, *3*(3), 653-661, 1967.

Hazen, Allen, Storage required for the regulation of streamflow, *Trans. Amer. Soc. Civil Eng.*, *77*(1308), 1914.

Herstein, I. N., *Topics in Algebra*, 342 pp., Blaisdell, New York, 1964.

Hufschmidt, M. M., and M. B Fiering, *Simulation Techniques for Water Resource Systems*, 212 pp., Harvard University Press, Cambridge, Massachusetts, 1966.

Hurst, H. E., Long-term storage capacities of reservoirs, *Trans. Amer. Soc. Civil Eng.*, *116*(776), 1951.

Hurst, H. E., Methods of using long-term storage in reservoirs, *Proc. Inst. Civil Eng.*, *5*(6059), 519-543, 1956.

Hurst, H. E., R. P. Black, and Y. M. Sunaika, *Long-Term Storage*, 145 pp., Constable, London, 1965.

Kendall, M. G., *A Course in Multivariate Analysis*, 185 pp., Charles Griffin, London, 1961.

Knuth, Donald E., *The Arts of Computer Programming*, vol. 2, *Seminumerical Algorithms*, 624 pp., Addison-Wesley, Reading, Massachusetts, 1969.

Luce, R. Duncan, and Howard Raiffa, *Games and Decisions*, 509 pp., John Wiley, New York, 1957.

Maass, Arthur, et al., *The Design of Water Resource Systems*, 620 pp., Harvard University Press, Cambridge, Massachusetts, 1962.

Mandelbrot, B. B., and J. R. Wallis, Computer experiments with fractional Gaussian noises, 1, Averages and variances, *Water Resour. Res.*, *5*(1), 228-241, 1969a.

Mandelbrot, B. B., and J. R. Wallis, Some long-run properties of geophysical records, *Water Resour. Res.*, *5*(2), 321-340, 1969b.

Mandelbrot, B. B., and J. R. Wallis, Robustness of the rescaled range $R/S$ in the measurement of noncyclic long run statistical dependence, *Water Resour. Res.*, *5*(5), 967-988, 1969c.

Matalas,  N. C., Statistical analysis of droughts, Ph.D. dissertation,

Harvard University, Cambridge, Massachusetts, 1958.

Matalas, N. C., Mathematical assessment of synthetic hydrology, *Water Resour. Res.*, *3*(4), 937-945, 1967.

Matalas, N. C., and Walter Langbein, The relative information of the mean, *J. Geophys. Res.*, *67*(9), 3441-3448, 1962.

Meramec River Basin Commission, The Meramec basin--Water and economic development, a report in 3 volumes, Washington University, St. Louis, Missouri, February 1962.

Meta Systems, Inc., Report to the State of Mississippi Research and Development Commission, Cambridge, Massachusetts, 1969a.

Meta Systems, Inc., A program for simulation of acid mine drainage in a river basin, report to the Appalachian Regional Committee, Cambridge, Massachusetts, 1969b.

Moran, P. A. P., *The Theory of Storage*, 111 pp., Methuen, London, 1959.

Noble, Benjamin, *Applied Linear Algebra*, 523 pp., Prentice-Hall, Englewood Cliffs, New Jersey, 1969.

Pisano, W. C., River basin simulation program, Federal Water Quality Administration, Clearinghouse no. AD673564, August 1968.

Raiffa, Howard, *Decision Analysis*, 309 pp., Addison-Wesley, Reading, Massachusetts, 1968.

Ralston, Anthony, *A First Course in Numerical Analysis*, 578 pp., McGraw-Hill, New York, 1965.

Rippl, W., The capacity of storage reservoirs for water supply, *Proc. Inst. Civil Eng.*, *71*(270), 1883.

Schaake, John C., Jr., and Myron B Fiering, Simulation of a national flood insurance fund, *Water Resour. Res.*, *3*(4), 913-929, 1967.

Sudler, Charles E., Storage required for the regulation of streamflow, *Trans. Amer. Soc. Civil Eng.*, *91*(1641), 622-660, 1927.

Thomas, H. A., Jr., Queueing theory of streamflow regulation applied to the cost-benefit analysis of a multipurpose reservoir, memorandum to the Harvard Water Program, Cambridge, Massachusetts, 1959.

Thomas, H. A., Jr., and Myron B Fiering, The nature of the storage yield function, in *Operations Research in Water Quality Management*, chap. 2, Harvard University Water Program, Cambridge, Massachusetts, 1963.

Thomas, H. A., Jr., and M. B Fiering, Fate of radioactive wastes in streams, 15 pp., report for the U.S. Atomic Energy Commission, Harvard University (Division of Engineering and Applied Physics), Cambridge, Massachusetts, 1964.

Tocher, K. D., *The Art of Simulation*, 184 pp., English Universities Press, London, 1963.

U.S. Army Corps of Engineers, Report of the Chief of Engineers on the Delaware River Basin, *House Doc. 522*, 87th Congress, 2nd session, Washington, D. C., 1962.

Wallis, J. R., and N. C. Matalas, Small sample properties of $H$ and $K$--Estimators of the Hurst coefficient $h$, *Water Resour. Res.*, *6*(6), 1583-1594, 1971.

Williams, J. D., *The Compleat Strategyst*, 234 pp., McGraw-Hill, New York, 1954.

Wilson, E. B., and M. M. Hilferty, Distribution of chi-square, *Proc. Nat. Acad. Sci.*, *17*, 684-688, 1931.

# NOTES

NOTES

# NOTES

# NOTES

# NOTES

**NOTES**